How Long is a Piece of String?

Other books you may enjoy:

What is a Googly? by Rob Eastaway
Why Do Buses Come in Threes? by Rob Eastaway and Jeremy Wyndham
The Hidden Mathematics of Sport by Rob Eastaway and John Haigh
100 Maddening Mindbending Puzzles by Rob Eastaway and David Wells

How Long is a Piece

ROB EASTAWAY AND JEREMY WYNDHAM

# of String?

ILLUSTRATIONS BY BARBARA SHORE

PORTICO

First published in the United Kingdom in 2021 by
Portico
43 Great Ormond Street
London
WC1N 3HZ

An imprint of Pavilion Books Company Ltd

ISBN 978-1-91162-226-0

A CIP catalogue record for this book is available from the British Library.

Illustrations by Barbara Shore

10 9 8 7 6 5 4 3 2 1

Reproduction by Rival Colour Ltd, UK
Printed and bound by 1010 Printing International Ltd, China

This book can be ordered direct from the publisher at www.pavilionbooks.com

CONTENTS

PREFACE TO THE NEW EDITION

Jeremy Wyndham and I wrote this book as a sequel to *Why Do Buses Come in Threes? Buses* had unexpectedly turned out to be a bestseller, to our delight.

We thought we had used up all of our material in that first book, but it didn't take long for us to find plenty more intriguing areas of everyday life where maths plays a part, and *How Long is a Piece of String?* was published in 2002.

Just one year later, Jeremy tragically died.

Among his many fine qualities, Jeremy was one of life's great optimists, but even he might have been surprised (as well as thrilled) that both of the books he co-authored would still be in print nearly twenty years later.

When Jeremy and I were first researching this book, we had the pleasure of encountering specialists from many different walks of life. These included Derek Smith, a lift guru at the company Otis, Richard Forsyth, who at the time was at Luton University and an expert on identifying the authorship of manuscripts, Pauline Matkins, then the fount of knowledge on taxi fares at Transport for London, and Matt Keeling, an epidemiologist at the University of Warwick. The chapter about epidemics has become more relevant than ever. At the time of the first edition, there had not been a viral pandemic for several decades. With the emergence of coronavirus, the mathematical modelling of epidemics became a mainstream topic.

For this new edition, I'm grateful to Ben Sparks, Zoe Griffiths, Kit Yates, Matt Chaffé, James Grime, Richard Harris, Jonathan Harris and Aoife Hunt, who have helped me to update those parts of the book that had become out of date.

I'm happy that Barbara Shore's original cartoons remain unchanged, and give the book the light, accessible touch that we wanted in a maths book aimed at a general audience.

Any errors that remain in the book are down to me, and to my failure to take heed of the Lincoln Index (see page 161).

INTRODUCTION

Imagine if your school timetable had offered the following optional topics.

Monday:	How to avoid being ripped off
Tuesday:	Thinking games
Wednesday:	Tips for highly paid jobs
Thursday:	Patterns in the real world
Friday:	When to take a chance

No doubt you would have chosen at least one of these options, and maybe all of them. And yet, without stretching reality too far, that is exactly how your timetable could have looked. It's just that some administrators decided to call each of these topics mathematics. Then, to be certain that all the fun was squeezed out, they made as much of the subject as abstract and detached from the real world as they could.

The result was that while a few children thrived, most spent hours ploughing through exercises that they found tedious and irrelevant. 'Miss, why do we need to know about Pythagoras?' 'Don't be impudent, Perkins.'

Fortunately, times have changed. Those wishing to engage the public in maths now appreciate that the way to do it is not to start with the theory, but to start with examples in the real world that touch people's lives. Much of maths is about abstract ideas, but for the vast majority of people it is accessible only if it can be understood in a context they are familiar with.

Today, for all sorts of reasons, it has become fashionable in some cultures to describe people as 'sad' for taking an interest in mathematics. Yet give us a topic that we care about, and we all become mathematicians. One of the most creative minds in history, Leonardo da Vinci, asked questions about everything he saw, and then investigated the answers. He was an artist, but arguably he was even more a scientist and a mathematician. As far as we know, nobody called Leonardo a nerd (or its Italian equivalent).

This is our second compilation of everyday maths. Once again, we have drawn from a broad range of topics that caught our interest. Our main criterion for including something has been that it kept us entertained in the pub. Some topics will be familiar to those well read in this field, but others – such as lifts, taxi fares and the logistics of the gentlemen's urinal – have had little if any public airing before now.

As with *Why Do Buses Come in Threes?* you will find some parts easy reading while others require more careful thought. As it happens, certain themes crop up more than once in the book. Chance, reasoning and patterns feature prominently. In fact if the syllabus at the start of this introduction really existed, perhaps this would be a book that accompanied it. But this isn't meant to be a textbook: it is to be read at leisure and, we hope, with enjoyment.

1

WHY DOES MONDAY COME ROUND SO QUICKLY?

How the numbers of the moon combined to make our week

In a book about everyday maths, what better place to start than the mathematics of the day itself, and of Monday in particular?

The seven-day week is so embedded in our culture that it is easy to forget that the idea of a week was just a convenient human invention. Why can't we really love somebody 'eight days a week' (to quote the song)? Or ten, for that matter? And why does the working week have to start on Monday?

In fact, like many things at the heart of our modern culture, the seven-day week came about thanks to a combination of superstition, coincidence, human error, a need for order – and some elementary mathematics. These things weren't just responsible for the week having seven days, they also determined the order of the days in the Western calendar. The reason why the week runs Monday, Tuesday, Wednesday, and not, for example, Wednesday, Monday, Tuesday, is to do with the way that numbers combine together.

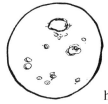 To see how the modern week evolved requires a rapid sweep through some of the story of the calendar, much of which, it should be pointed out, is still a source of debate among historians. The maths, however, is sound.

Where did the week come from? In the earliest tribes, there was no notion of a week mainly because it wasn't needed. The critical periods of time were the days and the seasons. For people, just as for animals, the day determined the basic survival routines of finding food, eating and sleeping. The seasons influenced the longer-term routines of hunting, harvesting and protecting against the climate.

Any tribe that could predict and plan for the different seasons would clearly have had better chances of surviving and prospering. Even the crudest of calendars would have given a tribe some advantage over rivals without such a tool.

So apart from the rough-and-ready clues of temperature and rain, how could ancient man have worked out what time of year it was? There is evidence that the first calendar was based on a very convenient clock in the skies – the moon.

The moon calendar – and the number twelve

After the sun, the moon is by far the most dominant presence in the sky, and it goes through an obvious cycle from one night to the next. A full moon slowly diminishes to a crescent and then no moon at all, and then returns to a full moon.

Archaeologists have found various clues that suggest that as far back as 30,000 BC the moon's cycle was being followed closely. Etchings found on bone resemble the moon in its various phases and there are scratches that tally closely with the days in each cycle.

There are good reasons why the moon would have been important to primitive people. The period of full moon to full moon coincides almost exactly with the period between ovulations for women. Whether any sort of family planning existed thousands of years ago we don't know, but the moon cycle, from which we derive the word menstruation, would at least have been a handy guide to fertility. Moon rituals, which still exist today, may have begun as fertility rites at that time.

There is another reason why the moon would have had such an attraction as a timekeeper. The cycles of the moon are a natural way of dividing up the year. There are roughly twelve moons in a year, so that became the obvious dividing number (the word 'month' also comes from the word 'moon', of course).

The exact number of moons in a year is, we now know, 12.36, and 12 just happens to be the nearest whole number. However, it takes quite a bit of rounding to trim 12.36 down to 12, and this was to prove a source of endless headaches to calendar makers from the Egyptians to Julius

---

Quiz: How big is the moon?

Which of the following circular objects when held at arm's length is about the same size as the moon?

(a) a pea
(b) a penny
(c) a ping-pong ball
(d) an orange

The answer is (a), a pea – a petit pois, in fact. For something that so dominates the night sky, this is surprisingly small. Thanks to the workings of the human brain, we perceive it to be much larger.

Caesar and beyond, as they sought to make the months coincide with the years. If the moon's orbit around the earth had been just a fraction faster, we would probably have ended up with a calendar of thirteen months, and as a result the number 13 might have become regarded as a *lucky* number because of its link with the year. But it was not to be.

If the fluke of there being twelve moons per year first established this number as a fundamental measure of time, then the discoveries of the early civilisations in Egypt and Greece set the number 12 in stone.

Twelve is a conveniently small number, and it has other useful properties. One of the most important is that the number can be divided into two equal parts, or three, or four or six – which makes it a practical quantity for measuring and sharing. Indeed the convenient divisibility of twelve helped it to survive in British culture in both the currency (twelve pennies in a shilling) and the main unit of length (twelve inches in a foot) until late in the twentieth century.

The number 12 has connections with a circle, too. One of the easiest ways of dividing up a circle is to use a pair of compasses.

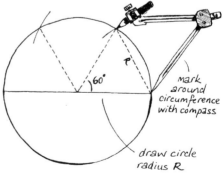

Marking around the circumference, the circle can be divided into six portions

60°

mark around circumference with compass

draw circle radius R

And once a sixth of a circle has been found, it is easy to divide it into halves:

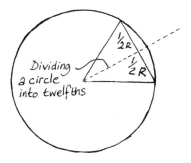

This means that a circle can very easily be divided into twelve equal pieces – very handy for dividing the sky into twelve portions to be symbolised by the signs of the zodiac, and later on, for dividing the clock face into hours.

The moon, the planets and the number seven

Every night, the familiar moon arches across the sky against a backdrop of twinkling stars. From the earliest times, it would have been observed that these stars were themselves slowly rotating, and that their rotation, like the sun's, took exactly a day. However, there were some exceptions. Some bright stars didn't follow the rest but moved at different speeds, and occasionally even looped back on themselves.

The small group of celestial objects with their own unique cycles acquired a special status. They became known as the wanderers, for which the Greek word was *planetes*. Since these planets each had their own distinct movements, it was only natural to give them names. By the time of the Romans, they were known in today's English as the Moon, the Sun, Jupiter, Saturn, Venus, Mars and Mercury.

12: an abundant number

All whole numbers have factors – that is, smaller whole numbers that divide exactly into them (except for 1, of course). The factors of 12 are 6, 4, 3, 2 and 1, which add up to 16. A number whose factors add up to more than the number itself is called 'abundant' and 12 is the smallest such number. Abundant numbers are in fact extremely common. Looking for something to get their teeth into, mathematicians therefore began to look not just at whether numbers are abundant or not, but how abundant they are. For example, 12's abundancy is 16/12 or 1.33. It is beaten by 24 (36/24 = 1.5) and this abundancy ratio increases with every multiple of 12 up to 60, whose factors add up to 108. (108/60 scores a whopping 1.8). The number 60 is highly abundant because so many numbers divide into it, and this Is why it became a popular base for counting. There seems to be no upper limit to how abundant a number can get, if you go high enough.

Does abundancy matter? Apart from enriching the understanding of numbers, no it doesn't. But ancient Greeks, particularly Pythagoras and his cronies, were always looking for evidence to support their view that numbers controlled the entire universe. Any obscure property of numbers took on a significance that might now be regarded as excessive.

The belief that there were seven so-called planets no doubt reinforced the mystical status of the number seven. But, as with the number twelve, it took another coincidence to seal the significance of the number seven in the calendar. It just so happens that the number seven is linked to the cycle of the moon. From full moon to no moon is about fourteen days – two lots of seven. Full moon back to full moon is just over 29 days – not far from four lots of seven, or 28. And 28 is also a number loaded with mathematical significance, as the box opposite shows.

This second occurrence of the number seven in the skies is a fluke, of course – the number of days in the moon cycle has

> ### 28: a perfect number
>
> *The factors of 28 are 1, 2, 4, 7 and 14 – and it so happens that these add up to 28. The Greeks spotted this coincidence, and labelled this number, and others with the same property, as 'perfect numbers'. Rather an arbitrary definition of perfection, you might think.*
>
> *Perfect numbers are rare. The Greeks discovered only four of them: 6, 28, 496 and 8128, and as far as we know they didn't find any others. This is hardly surprising since the next smallest perfect number is 33,550,336. It is believed that all perfect numbers end in 6 or 8, but it isn't known if there are infinitely many of them.*
>
> *The belief that numbers with curious mathematical properties were in some way mystical would certainly have helped to establish those numbers as part of the culture. The perfect number 28 and the abundant number 12 were two of the main beneficiaries.*

nothing to do with there being seven 'planets' – but, given mankind's natural tendency to read meaning into any coincidence, it is hardly surprising that early civilisations believed that the number seven had a fundamental link with the heavens. For convenience and for ritual, the moon cycle was therefore divided into partitions of four lots of seven days.

'In the beginning', according to the Bible, God took seven days to create the earth. Is this an even more fundamental reason for the mystical importance of the number seven? Or did the people who wrote this story choose a number that already carried significance for them because it divided up the lunar month and represented the number of the planets? It doesn't really matter. The fact is, this story firmly established the idea of a seven-day week in Jewish culture with the seventh day being named the rest day, or Sabbath. The Jews spread the idea across the Middle East, and the Romans adopted it, too, despite the fact that they already had a separate eight-day 'market' week of their own whose origin is obscure.

Florence's symbol of the eight-day week

In medieval Florence, they believed in an eight-day week – of sorts. In front of Florence cathedral is an octagonal building, the Baptistry. The shape of the building has a significance. Seven of its eight sides represent the seven days of the earthly week. The eighth side of the octagon represents the eighth, eternal day, the day we spend in heaven (or the other place) after we die.

It also happens to be a lot easier to build an eight-sided building than a seven-sided building, so the eight-day symbolism got the architect out of an awkward design problem.

Linking the hours and the planets

We've seen how, quite independently, the numbers 12 and 7 became crucial to the measurement of time, 12 because it divided up the year and, later, the day, and 7 because it divided up the month. The modern week was now ready to evolve, and the names of the days were to emerge because of one further link between those two mystical numbers.

It had been established early in astronomical history that each of the planets took a different amount of time to complete a cycle and return to its starting position – one planet 'year'. This led the planets to be given a hierarchy, with Saturn, the planet with the longest cycle, being the most senior. The full hierarchy follows at the top of the next page.

Now you might expect that, with seven planets and seven days, it would be a logical and simple step just to name the days

Saturn (29 years)
Jupiter (12 years)
Mars (687 days)
Sun (365 days)
Venus (225 days)
Mercury (88 days)
Moon (28 days)

in the order of the planets: Saturnday, Jupiterday, Marsday and so on. But, by a quirk of astrological practice that is still not fully understood, this isn't what happened.

The Egyptians were the first to divide the daylight into twelve hours, and then, around 1000 BC, the Babylonians, based in the region around modern-day Iraq, divided the day and night into 24 hours. Instead of naming the days after the planets, they decided to name the *hours* after the planets. The first hour was allocated to the most senior planet, Saturn, the second hour to the second planet, Jupiter, the third to Mars, and so on, repeating the cycle of the seven planets throughout the 24 hours and continuing it into the following days. The result looks like this (read down the columns and continue the cycle from one column to the next):

HOUR	DAY 1	DAY 2	DAY 3	DAY 4	DAY 5	DAY 6	DAY 7	DAY 8
1	SATURN	SUN	MOON	MARS	MERCURY	JUPITER	VENUS	SATURN
2	JUPITER	VENUS	SATURN	SUN	MOON	MARS	MERCURY	JUPITER
3	MARS	MERCURY	JUPITER	VENUS	SATURN	SUN	MOON	MARS
4	SUN	MOON	MARS	MERCURY	JUPITER	VENUS	SATURN	SUN
5	VENUS	SATURN	SUN	MOON	MARS	MERCURY	JUPITER	VENUS
6	MERCURY	JUPITER	VENUS	SATURN	SUN	MOON	MARS	MERCURY
7	MOON	...etc...	...etc...	...etc...	...etc...	...etc...	.. etc...	...etc...
8	SATURN							
9	JUPITER							
10	MARS							
...etc								
22	SATURN	SUN	MOON	MARS	MERCURY	JUPITER	VENUS	SATURN
23	JUPITER	VENUS	SATURN	SUN	MOON	MARS	MERCURY	JUPITER
24	MARS	MERCURY	JUPITER	VENUS	SATURN	SUN	MOON	MARS

Because 7 doesn't divide exactly into 24, the planet at the top of each day changes. In fact, because 24 divided by 7 has a remainder of 3, the planet at the top of the list skips three planets down the planet hierarchy each day. On Day Two the Sun tops the list, on Day Three the Moon, and so on. After seven days each of the seven planets has been top of the list, and on Day Eight the cycle begins again.

The planet at the top of each day became known as its 'dominant' planet, and it became the custom to name the day after this dominant planet. So we have:

'DOMINANT PLANET' ON EACH DAY	
DAY 1	SATURN DAY
DAY 2	SUN DAY
DAY 3	MOON DAY
DAY 4	MARS DAY
DAY 5	MERCURY DAY
DAY 6	JUPITER DAY
DAY 7	VENUS DAY

Sound familiar? To make it explicit, here is this planet week set against the modern English and French weeks:

'DOMINANT PLANET' DAYS	THE ENGLISH WEEK	THE FRENCH WEEK
SATURN DAY	SATURDAY	SAMEDI
SUN DAY	SUNDAY	DIMANCHE
MOON DAY	MONDAY	LUNDI
MARS DAY	TUESDAY	MARDI
MERCURY DAY	WEDNESDAY	MERCREDI
JUPITER DAY	THURSDAY	JEUDI
VENUS DAY	FRIDAY	VENDREDI

The days highlighted in bold have retained their planet names. And remember that this order has been determined by the fact that 24 divided by 7 has a remainder of 3.

This seven-day planet week eventually reached the Romans, who spread it across their empire, and it became the norm throughout Europe, with one small amendment. When Christianity took hold in the Roman Empire in the fourth century AD, the Romans felt it important to symbolise the change by making their week distinguishable from that of other religions. Since the Jews' most holy day was Saturday (the Sabbath), the Christians appointed another day – Sunday – as their day of rest. They scrapped the pagan Sun god and renamed it 'The Day of the Lord' *(Dies Dominici)*. Countries closest to the centre of the Roman Empire were the likeliest to adopt this, and, in many European countries today, *Dies Dominici* survives in a slightly corrupted form *(Domenica* in Italy, *Dimanche* in France, *Domingo* in Spain). Apart from this change, these romance languages have stuck with the planets, and always in the Babylonian order. Which means, of course, that the first working day of the week became Moon-day, the day after the rest day.

The Britons, who were that much further away from Rome's religious influence, preserved the Sun-day, but sacrificed the last four planet days to Anglo-Saxon gods, Tiw, Woden, Thor and Frig. Well, when you've just been invaded and had your village pillaged, you don't have much choice.

The time of day, the weeks, months and year are a great reminder of how numbers and mathematics are at the very root of our culture. The seven-day week owes a lot – perhaps everything – to early civilisations counting what they believed were seven planets.

How different things would have been if Neptune, Pluto and Uranus had been close enough to be visible to the naked eye. People of the Stone Age would have counted ten planets – one for each finger – which would surely have put ten beyond any question at the top of the number tree and would no doubt have led us to the ten-day week and the three-week month. Think of the consequences. On the downside, it might have meant more work and less play. But at least we would have 30 per cent fewer Monday mornings.

2

HOW DO CON ARTISTS GET RICH?

Tricks and scams that make people part with their money

A woman buys a £100 ring from a jeweller, and just as she is about to leave the shop she pauses and returns to the counter.

Woman: I'm not happy with this ring. Can I exchange it for that £200 one there?

Jeweller: Certainly, madam (handing it over to her.) That will be another £100, please.

Woman: Excuse me, but I don't owe you anything. I handed over £100 just now, and on top of that I've just handed you a ring worth £100. That's £200 altogether.

And with that she storms out of the shop clutching the £200 ring, leaving the jeweller to ponder where he went wrong.

It's easy to sympathise with the confused jeweller. You need quick wits to avoid being conned. No wonder when exposed to more sophisticated scams, so many people are taken in.

Predicting the sex of your baby

Take the following (fictitious) advertisement, for example:

BABY-SEXING (THE NON-INTRUSIVE,
TOUCH METHOD)
*Do you want to know the sex of your baby, but don't like
the thought of having sound waves pumped into your
womb? We have a completely non-intrusive method that
will determine the sex of your baby using only human
touch. The price is just £100. In the event of our getting
it wrong, you not only get all your money back, but also
an additional £50 in compensation. Phone Jackie on ...*

Sounds like a reasonable offer. After all, there aren't many
services that promise a full refund *and* compensation if they
get it wrong. Why can't more services be as straightforward
and honest as this one?

But, if you buy into this deal, then you've been had. It turns
out that the only skill Jackie needs in order to offer this service
is the ability to toss a coin. Heads is a boy, tails is a girl. Before
she visits you, she tosses a coin to decide on the sex of the baby.
She goes through her hand-sensing ritual, declares, 'It's a boy'
and then pockets your £100. About half the time, of course, she
is right, and keeps the money. The
other half of the time she is wrong,
so she has to pay you £150. Well no:
actually, it's only £50, because the
other £100 was yours in the first
place.

So if Jackie performs her
baby-sexing service 100 times,
then on average:

- she will make £100 fifty times (total income £5,000)
- she will lose £50 fifty times (total loss £2,500)

In other words, with 100 customers she can expect to make a profit of about £2,500, which works out at £25 a go. Just for tossing a coin!

This scam, similar to many that have been perpetrated, is probably illegal on the grounds of obtaining property by deception. Is it harmful? Yes, in that it is preying on vulnerable people, and conning them out of £100 that they could more usefully have spent on baby clothes.

Like many cons, it is founded on an initial offer that sounds too good to refuse.

How to show somebody that full is empty and empty is full

It is often said that an optimist describes this glass as half full, and a pessimist says it is half empty. Of course we know that they are the same, thing. And, if two things are equal, we can put them in an equation:

$$\text{Half full} = \text{Half empty}$$

Now abbreviate this into letters:

$$\tfrac{1}{2}F = \tfrac{1}{2}E$$

With an equation, if you double one side, you need to double the other.

$$2x(\tfrac{1}{2}F) = 2x(\tfrac{1}{2}E)$$

Cancelling out we get:

$$F = E$$

In other words Full = Empty.

The football con

George Tindle was in the middle of his morning routine of deleting junk emails when he noticed one that grabbed his attention. It was headed AMAZING FA CUP PREDICTION. Curiosity aroused, he clicked on it to find out more. He found this message:

Dear Football Fan

We know you will be sceptical, but we have devised a remarkably accurate method of predicting the results of football matches. This afternoon, Burnley play Derby County in the third round of the FA Cup. Our system forecasts that Burnley will win. We advise you not to bet on this, but you might be interested to note the result this afternoon.

Yours truly,

The Cup Predictors

George gave a little smile, but didn't give it any further thought until later that day when he tuned in to watch the results, as he always did. Burnley won their game. 'Probably the favourites in any case,' he thought.

Three weeks later, another email arrived in the box.

Dear Football Fan

Do you recall that we accurately predicted the victory of Burnley in the last round of the FA Cup? Today, Burnley plays Middlesbrough. Our prediction is that it is Middlesbrough who will go through to the fifth round. We strongly advise you not to bet, but please follow our results to see if they are correct.

Yours truly,

The Cup Predictors

George was a little curious, and awaited the results that afternoon with just a bit more interest. Result – a 1-1 draw. See – it had all been a fluke.

But the following Tuesday, Middlesbrough won the replay 2-0. The Cup Predictor email followed days later. This time it forecast an upset in the fifth round, Blackburn Rovers would defeat Middlesbrough. And they did. And so to the quarter-finals, when Blackburn were forecast to lose to Tottenham. Correct again. That was four out of four.

'We know this is an unusual system,' said the next email, 'but maybe you are now more convinced that we are on to something. In the semifinal, Arsenal will beat Everton.' George couldn't believe it. He'd already told lots of his friends, and together they followed the live commentary that afternoon. Despite going behind, Arsenal went on to win 2-1. This was incredible.

The next day, another email arrived.

Dear Football Fan

You have witnessed the results of our amazing football predictor system. Are you convinced? We have made five correct predictions out of five, which you will agree defies normal chance, especially since the team that won was not always the favourite. As a special deal, we are offering you a chance to subscribe to a trial month with our match forecasting service for just £200. You email us the two teams and we send you the prediction.

We look forward to receiving your order.

Sincerely Yours,

The Cup Predictors

'Two hundred pounds is a bit steep,' thought George, 'but, then again, if I know who's going to win I'll be able to make that money back from the bookmaker a thousand times over.'

He got out his credit card. And he had swallowed, hook, line and sinker.

But where was the con? Unlike the baby predictor, here we had five correct predictions. Surely there had to be something in it. To see how the con worked, let's look at some of the other customers who received emails from the Cup Predictors.

In an office down the road, Jim received an email on the first morning of the scam, just like George. But while George's email said 'We predict that Burnley will beat Derby County', Jim's email was curiously different. It said 'We predict that *Derby County* will beat *Burnley*'. When Derby County lost, Jim didn't receive any more emails. Ten miles away, Debbie was given Burnley as her first prediction and her second prediction. When they lost to Middlesbrough in Round 4, her emails stopped, too.

In fact, the scam was deceptively simple. To start with, 8,000 emails were sent out to people known to have some sort of interest in football. A match was chosen at random, and half the recipients were told Burnley would win, the other half Derby County. Of course, 4,000 would be 'right', while the other half would delete the email and think no more about it.

In the next round, 2,000 got the prediction of Burnley, and 2,000 Middlesbrough. After the result of this, 2,000 would be certain to have received two correct predictions out of two. Of course, the Cup Predictors continued sending emails only to the winners, so that, by the time of the final, 250 people had received five correct predictions. And those 250 people felt very special. (Wouldn't you ?) So special that 50 of them handed over the £200. This gave a nice profit to the organisers, who had done nothing at all but send out emails.

Scams like this depend on our natural inclination to think we are special, so that when lucky things happen to us there must be a reason. Out of every 32 people receiving predictions, only one of those is guaranteed to get all five correct. The other 31 will receive an incorrect prediction, at which point the messages stop. It just so happens that on this occasion it is George who is the lucky one, and so of course he feels special. But remember, like the National Lottery jackpot, it had to happen to somebody.

Like the baby con, this football story is fictitious, but illegal scams similar to this one have been operated many times. You can imagine that this kind of thing is particularly successful in the pseudo-scientific world of the stock market, where a dodgy adviser might advise half his prospective customers that a certain share will go up, and the other half down*.

---

* In 2008, the illusionist Derren Brown pulled off a similar scam in which he convinced members of the public that he had found a way of predicting winners of horse races. The programme was called *The System*.

The famous restaurant con

Three men in a restaurant are presented with a bill for £25. They give the waiter £30, in £10 notes, and he returns with £5 in loose change. The men take £3, and leave £2 as a tip.

The men have now each paid out £9, making £27 in total. The waiter has a £2 tip. £27 + £2 = £29. But the men gave the waiter £30, so £1 is missing. Who is conning whom?

(See the end of the chapter.)

The pyramid problem

One of the most successful and harmful scams of all comes under the general heading of 'pyramid selling'. Unlike the earlier schemes in this chapter, pyramid selling does genuinely offer the public the chance to make money – but only if they can rope other people in to be the losers. And, in some forms and in many countries, pyramid schemes are still legal.

One famous example of pyramid selling was a scheme called 'Women Empowering Women', which has appeared in various guises over the years. This scheme pulled on a powerful emotional heartstring. It claimed that most moneymaking schemes are run for and by men. Here at last was a chance for the sisterhood to make money for itself. No men would be involved. The idea struck a chord with women who felt that men had been exploiting them for far too long. Little did they realise that they were going to be exploited by women instead.

The system was extremely simple. To join cost £3,000, though this £3,000 wasn't described as a cost, it was an 'investment'. The joiner got to put her name on a heart chart. She would then be able to recruit other 'hearts' to join under hers, each of whom had to 'invest' £3,000 in her. As soon as there were eight hearts under her name, she left the scheme, having received the £3,000 from each of her eight supporters, making £24,000 in total. Since she had paid £3,000 to join the scheme, she had made a net profit of £21,000.

And the great thing for many women is that it worked: £3,000 had turned into £24,000.

However, it is mathematically impossible for this scheme to work for everybody. After all, nothing is being produced. All that is happening is that some women were transferring £3,000 to other women. For every person gaining £21,000, there had to be seven who were losing £3,000.

In theory, in an infinite population, a pyramid scheme like this could go on for ever. Imagine if you have just bought into it. With enough skill of persuasion, you could certainly find eight people, maybe friends, who were prepared to part with £3,000. You could reassure them with the fact that it has worked for you and everyone above you in the pyramid, so it can work for them, too. So long as the scheme is in its early days, you are likely to succeed.

But, perhaps without realising it, your claim that 'it worked for me so it will work for you' is a lie. The population is not infinite. Eventually – maybe when a thousand are in the scheme, or maybe when there are a million – the number of people able or prepared to join will begin to dry up. Either they will already be members, or they will not be willing to part with £3,000. At that point, the whole scheme will fold. And, when the scheme folds, anything up to seven-eighths, or 87.5 per cent, of the people who entered the scheme will discover that they have spent £3,000 that they will never see again.

It's a bit like a game of pass the parcel, except in this version everyone left holding the parcel when the music stops is a loser.

Pyramid schemes all work in a similar way to this. They don't have a product, all they offer to do is make you money, which will come from recruiting other people. And the idea is so powerful, it can even bring down a whole economy.

How pyramid selling nearly ruined Albania

In 1996, the country of Albania was brought to its knees by a pyramid scheme. The country was emerging from communism, and most people were living in poverty. The people were therefore particularly vulnerable to a scheme that offered to make them rich quickly. The fact that the scheme seemed to be supported by the country's major banks and its government just fuelled the enthusiasm even more.

In the case of Albania, what was being offered was a fantastic interest rate on money invested. Everybody sees special interest-rate offers from banks from time to time, but what should have aroused the suspicion of investors was that the schemes were offering far higher interest rates to investors than banks were charging for lending money. This couldn't be right. Think about any loan or mortgage that you take out. The interest you pay is going to be higher than the interest rate on the money you keep in the building society. That's how moneylenders make their money.

So how were the Albanian pyramid lenders able to offer such high returns? They did it by using the money deposited with them by lenders to pay the interest.

To explain why this was bound to end in ruin, let's set up a scheme that is a simplified version of what happened in Albania. We'll call the scheme 'Sphinx Investments'. Here's the deal.

Sphinx are asking you to invest £100 in their amazing new scheme. In return, they will pay you 25 per cent interest every year. In other words, you can cream off £25 each year as long as you leave your £100 invested. What this means is that in four years you will have doubled your money. Fantastic, especially when compared with a building society, where £100 deposit might only earn you £1 a year.

What you don't know is that, when Sphinx start this scheme, they have absolutely no money in the bank. They are going to be using your money to pay your interest. And, for a while, they can do this – as long as their generous interest rate brings in a steady flow of new investors.

Attracted by the prospect of earning 25 per cent in interest, in the first year 1,000 people decide to join the Sphinx scheme and pay in their £100. So by year end, Sphinx have £100,000 in their account. They do have to pay out interest, though, of £25,000. But Sphinx can pay themselves a nice commission of 10 per cent and still end the year with £65,000 in their bank account:

A	B	C	D	B – C – D
Number of new investors	Money deposited by new investors (£100 each)	Interest paid out at year end	Sphinx's Commission (10%)	Amount in Sphinx bank account
1000	£100,000	£25,000	£10,000	£65,000

The next year another 1,000 join and invest their £100,000. At the end of the year, Sphinx have to pay 25 per cent to all their investors (this year's and last year's), so the interest payments have gone up to £50,000. But Sphinx still end up with £105,000 cash in the bank, which is well up on last year.

A	B	C	D	E	B+C−D−E
Number of new investors	Money deposited by new investors (£100 each)	Amount in bank at end of previous year	Interest paid out at year end (25% of total deposits)	Sphinx's commission (10%)	Amount in Sphinx bank account
1000	£100,000	£65,000	£50,000	£10,000	£105,000

So Sphinx seem to be making money out of offering good interest, and the customers are certainly making good money. Those who invested in the first year have already received £50 in interest on their £100 investment, and, not surprisingly, they are telling their friends all about it.

But, as with all pyramid schemes, these short-term gains are building to a disaster. If 1,000 people keep joining each year, this is what happens. The critical column to keep an eye on is the one on the right:

End of year	Interest offered	Number of new investors	Money from new investors (£100 each)	Total deposits so far	Interest paid out at year end	Commission each year	Cash in Sphinx bank account
1	25%	1000	£100,000	£100,000	£25,000	£10,000	£65,000
2	25%	1000	£100,000	£200,000	£50,000	£10,000	£105,000
3	25%	1000	£100,000	£300,000	£75,000	£10,000	£120,000
4	25%	1000	£100,000	£400,000	£100,000	£10,000	£110,000
5	25%	1000	£100,000	£500,000	£125,000	£10,000	£75,000
6	25%	1000	£100,000	£600,000	£150,000	£10,000	£15,000
7	25%	1000	£100,000	£700,000	£175,000	£10,000	−£70,000

At the end of Year 4, the amount being paid in interest and commission (£110,000) is more than the amount of money being deposited by new customers (£100,000). As a result,

Sphinx's bank balance begins to drop for the first time. By Year 6, cash reserves have plunged to just £15,000 and, the next year, Sphinx are in debt – so they can no longer pay interest. But this isn't the worst of it. As soon as they get a hint of financial problems, the investors all decide that now is a good time to take out their £100 deposit. To their horror, they discover that Sphinx have no assets, and so their £100 is gone for ever.

As with all pyramid schemes, a minority – those who were in at the start and who spread the word about this 'wonderful scheme' – have made a profit. In fact those who entered at the start of Year 1 have made £150 interest at the end of Year 6, so they still made 50 per cent gain even if they lost their deposit. But the vast majority have lost money.

Sphinx's problems were caused by bad cash flow, that is, running out of money with which to pay customers. They could have improved things if the number of new customers per year had grown. But, as with 'Women Empowering Women', customer numbers cannot grow for ever. They are only delaying the inevitable.

What happened with Sphinx is much the same as what happened in Albania, though the Albanian scheme was made worse because the government was duped into promoting the scheme and preyed on the poor, who could not afford to lose their money. In its final days, the scheme was able to keep attracting new investors only by increasing the interest rates to ludicrous amounts, which made the final crash even more rapid and disastrous. The damage to the economy was huge, and its impact is still being felt today.

All bubbles burst eventually

Pyramid schemes are self-created bubbles, which continue to inflate until the money runs out or investors lose their nerve.

Because they are built on nothing but promises, they cannot make money for anyone except for the people who set them up and the early investors.

Bubbles like this aren't always the result of fraudulent schemes. The dotcom explosion and various booms in the property market have been caused, at least in part, by a mass of people jumping on to a bandwagon because they saw a few people get rich quickly. After a while, people buy purely on the assumption that they will be able to flog what they have to somebody else for even more. It has to stop somewhere.

But as these schemes all demonstrate, by the time a 'good earner' becomes well known, it's probably already too late to make easy money. And like gambling, the books have to add up. If some people win, it's only because others lose just as much.

The restaurant con

Remember the three customers in the restaurant earlier? Well, it is the authors who are conning any reader who believes £1 really is missing. All that is being presented here is some misleading accounting.

At the end of the transaction, the men have paid £27, of which £25 was for the meal and £2 for the tip. £27 -£2 = £25, which balances perfectly.

Another way to look at it is that the men paid £30, of which £25 went to the meal, £3 returned as change, and £2 went on the tip.

The sum £27 + £2 = £29 is a complete red herring, but because it is so close to £30 it is easy to believe the two numbers are related.

3

WHAT MAKES A HIT SINGLE?

Patterns and variations that everyone likes

Of the millions of tunes and songs that have been written, a few are destined to grab the attention of a whole generation. In modern culture, the big hits tend to belong to a particular type of popular music – songs performed most often by attractive singers, and usually about love or relationships. But every era has had its own form of pop music. Some music now regarded as 'classical' would have been the people's music of its day, and societies have always had favourite folk songs as an important part of their culture.

Are there any rules to say what will make a tune a hit? Recording companies would love to know the answer to this question. In a way, they have already answered part of it. Hence the phenomenon of manufacturing pop groups rather than allowing them to form by chance.

But, setting aside the obvious attractions of sex symbols or topical words, what else are the essentials for giving a tune mass appeal?

Harmonies and melodies that fit our understanding of notes in a scale are clearly one essential, and these are discussed in

Chapter 14. But there are other elements to do with numbers and patterns that are arguably even more basic.

Everyone's got rhythm

What is it about a drumbeat that makes it so important in popular music? There is one obvious explanation. Built into us we have our own thumping drum, in the form of a heart. Typically it beats seventy times in a minute. Amid the sloshing and muffled sounds that we were all exposed to in the womb, one dominant sound would have been the beat of our mother's heart. So it would be surprising if in later life the beat of a heavy drum didn't have all sorts of associations with security and the world around us.

In popular songs it is common for the heavy drumbeat to occur at roughly the same rate as a heartbeat. That beat will vary considerably from song to song, just as a heart varies from slow to fast. Fast drums are associated with excitement or youth (sometimes both), where the heart also beats faster. A fast drumbeat of, say, 110–120 per minute usually creates a feel-good, sometimes even a high-adrenaline, mood, and that rate is typical of a Lady Gaga song, for example.

Music doesn't just work in single beats, however. In most tunes, the underlying rhythm involves a loud beat followed by one or more quieter beats.

The simplest such rhythm is the march. The British Army of old loved to march while whistling tunes with a two-beat rhythm like 'The British Grenadiers' or 'Colonel Bogey' (the latter most famous as the theme to the film *Bridge on the River Kwai*). Whistle one of these tunes and the simple left-right-left-right rhythm is obvious.

Almost as common is a three-beat. Once again, this is associated with foot movement, though in the more leisurely form of a waltz. Again, the 'triangular' nature of waltz tunes becomes obvious pretty quickly if you hum to yourself a tune such as 'The Blue Danube', or the song 'Close Every Door to Me' from the musical *Joseph and the Amazing Technicolor Dreamcoat:*

Close	Ev'	ry
1	2	3
door	to	me
1	2	3
Hide	all	the
1	2	3
world	from	me
1	2	3

… and so on.

The most common beat in popular music is based on the number four, whether it is the beats in rock and roll or the rhythm of a quickstep.

The Beatles were fond of counting-in their songs with 'One, two, three, four' before the opening chord. Four beats are really two lots of two, of course, and in rock and roll it is normal for the basic drum to beat on every second note.

Although tunes do occasionally come in fives – the theme tune to *Mission Impossible* and a jazz piece by Dave Brubeck called 'Take Five', for example – just about every popular tune will be composed of basic rhythms of twos or threes.

Why don't songs with five beats usually make it to Number 1? Almost certainly it is because of the patterns that our brains are wired to recognise. Investigations into the brain in recent years have established that whatever our mathematical ability, almost all of us are preprogrammed to be able to recognise patterns of one, two or three before we learn any arithmetic.

The experiments used by Karen Wynn to test this out received quite a bit of publicity. Babies just a few months old were shown a fluffy doll on a small stage. A screen was then raised in front of the stage. The experimenter had a secret hole through which she could secretly add or remove dolls. When the screen was lowered again, the baby lost interest if there was still just the one doll inside, but paid more attention if a second doll had been added. This demonstrated that a baby knew that one is not the same as two.

Further experiments showed that the baby showed added curiosity if 1 + 1 produced only one toy, since it expected one doll plus one doll to be two dolls.

In fact, using this and other experiments, it was established that most of us already knew 1+1= 2, 2-1 = 1, and 2+1 = 3 before we could speak. After 3, however, our instinct becomes less reliable.

In the same way, it is likely that our brains can recognise beats of one, two and three without thinking, so that we are drawn by instinct to these rhythms and their multiples. When listening to rhythm we are subconsciously counting, and for patterns of twos and threes this counting is effortless. That isn't to say we don't like rhythms of five, say, but we are less automatically drawn to them. Popular tunes use simple numbers.

The Mozart effect

In 1993, Nature magazine published an article entitled 'Music and spatial task performance'. It reported that listening to a Mozart piano sonata for ten minutes could improve various problem-solving skills for up to a quarter of an hour afterwards.

The idea that music could make us more intelligent captured the popular imagination, and became known as 'the Mozart effect'. Today it is common for expecting parents to play Mozart (or reggae or anything else they find inspiring) for the foetus to listen to. The hope is that this will give their child's brain development a head start. Elderly people often claim that solving puzzles keeps their brain supple. Perhaps certain types of music serve the same purpose.

The Mozart effect on its own is unlikely to be the key to turning anyone into a maths genius. Nevertheless, it is yet another example to add to the centuries-old connections between music and mathematics. It's worth noting that there is plenty of evidence that Mozart himself, like many musicians, had a keen interest in numbers. For example, in the margin of one of his fugues he scribbled calculations relating to his chances of winning a lottery.

The importance of variety

One of the secrets of popular patterns, especially songs, is their predictability. The regular beat, the familiar chords and the formulaic pattern of verse, chorus, verse, chorus mean that the mind is not too challenged, and listening is easy. By the way, this applies just as much to hymns and many classical tunes

Why even numbers are funkier than odd numbers

If you've ever run a stick along a fence, you'll be familiar with the regular 'ra-ta-ta-ta' noise that it produces. But if some of the fence posts are missing, it's possible to detect some well-known rhythms. For example, if you remove posts 2, 3, 6 and 8 from an eight-post fence (not that you are being encouraged to vandalise, you understand) the fence ends up like this:

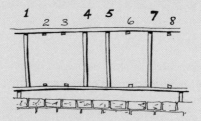

Run a stick along it at an even speed, and you should hear the rhythm of 'MAN ... u-NI-TED' or 'NOR-thern IRE-LAND' in the way that football crowds (used to) chant it. This pattern is unique – no other removal of posts quite fits. Interestingly the same rhythm has completely different meaning in other contexts. 1 x x 4 5 x 7 x is also the basic rhythm of a tango.

You only have to change the tango subtly to get something very different 1 x x 4 x 6 x 8 is the rhythm behind the theme to The Simpsons.

You can reproduce the opening sequence of that theme by removing posts from a 32-post fence. It goes 1 x x 4 x 6 x 8 9 x x 12 x 14 x 16 17 18 19 20 x x x x x 26 27 28 29 30 x x

In this Simpson rhythm, notice how there are more even numbers than odd numbers. Rhythms with lots of even numbers tend to be funky, jazzy or Latin influenced. In other 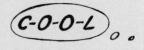 *words even numbers are often cooler than odd numbers.*

as it does to pop music. The great composers like Beethoven and Tchaikovsky all followed well-understood rules about the structure of a symphony.

However, unless the aim is to enter a trance-like state, nobody likes music to be too repetitive. Music that is totally predictable soon becomes boring because it doesn't require us to think.

There is more than one way to create patterns within the acceptable rules, and talented musicians make a name for themselves by both challenging the boundaries and yet still staying within the rules.

Mozart was famous for this. Musicians think of Mozart as a genius for his ability to produce music that is easy to listen to yet full of clever surprises at the same time. Mozart was playing with us, exploring some of the alternative patterns and sequences that are possible within music.

It is difficult to appreciate Mozart's approach without listening to some music, but something similar can be demonstrated with a little non-musical experiment.

Here is a question about a simple pattern: If ABC goes to ABD, what would you say XYZ goes to? Think about your answer before reading on.

Did you come up with XYA? If you did, you are in agreement with at least 80 per cent of the population. XYA fits a familiar pattern. What comes after Z? A does, because it starts the cycle again. This is a bit like a composer finishing a tune with a solid, comfortable major chord.

But XYA is not the only answer. There are many other possible patterns, depending on the rule for how to choose the symbol that comes after Z. For example, on a computer spreadsheet, the column after Z is AA. This makes XYAA a possible answer. Maybe, though, the rule is that numbers follow letters, which leads to XY1. Or perhaps *nothing* comes after Z, which means the answer is XY. What other possibilities can you come up with?

Some of the possible answers may strike you as pretty, while others may seem unsatisfying. This sense of a pattern feeling 'pretty' or 'satisfying' is similar to the effect that different endings to a musical piece might create.

What would a Mozartian answer be? Perhaps it would be WYZ. WYZ is an imaginative answer to the original puzzle. If Z can't go outwards, then X must go inwards. It is clever and symmetrical, and yet few people think of it. Just the sort of effect that Mozart liked to create. Maybe you can think of popular artists of today who achieve the same sort of thing.

Finding the right balance

A critic once said that his idea of hell was music that is totally predictable, or music that is totally unpredictable. He summed up what many of us instinctively know. Almost certainly, for a song or a tune to become a hit it needs to find the right balance. We've already said that too little variety makes a tune dull, but too much variety makes a tune impossible to follow. The ultimate in variety would be a tune composed entirely of randomly selected notes.

The predictability of music can actually be measured. By sampling at regular intervals, it is possible to measure the degree of predictability of the successive notes in a tune. If you think of a tune that is just a repetition of the note middle C, every sample would be exactly the same as its predecessor. The correlation in this music would be 100 per cent. In contrast, if you choose each note by rolling a die that is numbered 1 to 88 to represent every note on the piano keyboard, each note will have no predictable connection with its predecessor, so the correlation will be close to zero.

Will they ever run out of tunes?

Hundreds of new songs are published every week, but how many more new tunes can there be? The supply of tunes in Western music is limited by the number of permutations of the twelve-note scale. It is limited further by the fact that a large proportion of the possible sequences of notes don't sound good, and are inappropriate for popular tunes in our current culture.

However, even if we are limited to only a certain combination of notes that go together well to make a tune, the scope for variety is still enormous. Denys Parsons, a researcher, discovered that it was possible to identify melodies by noting whether each successive note was higher (U for up), lower (D for down) or the same (R for repeat) relative to its predecessor. Take the tune for 'Happy Birthday', for example. The second note is the same as the first (R) the third goes up (U) the fourth goes down (D).

In fact the tune goes: R U D U D D R U D U D D R U D ... and so on, but there is usually no need to go beyond fifteen letters in the sequence. That pattern of Rs, Us and Ds is unique to 'Happy Birthday' among all popular tunes. In fact, this should not be too surprising. There are 3^{15}, about 14 million, different ways of mixing R, U and D in a sequence of fifteen. So it would be possible to produce five hundred tunes per week for more than five hundred years and still come up with new R, U, D patterns in the first fifteen notes. And this is before we add in other variations on top: the notes themselves and the length of time between the notes (the rhythm), will also create new tunes.

On that evidence, the music industry is here to stay.

Music with a high level of correlation is known as *brown**
music, while music that is highly random is called *white* music.
This latter is related to the term 'white noise', which is the
random cracklings that can be heard on a radio when the dial
is between stations. The music in between white and brown –
predictable but not too predictable – is known as pink music.

In the realm of published music, analysis by Richard Voss
and John Clarke in 1975 suggested that all popular tunes fall
within the pink band. Tunes by minimalist composers such as
Philip Glass would presumably rate at the very brown end of
pink. Mike Oldfield's *Tubular Bells* would be slightly pinker
but still on the brown side. The sound of an orchestra tuning up
would be at the white end of pink. But it seems that the most
popular music – from Ella Fitzgerald to Michael Jackson – sits
firmly in the middle of the *pink*** zone.

So could there be a mathematical formula for producing pop
songs? Well to some extent it is already happening. A variety of
companies use Artificial Intelligence (AI) techniques to create
new tunes to any particular style that you care to choose. They
do it by patching together musical phrases and patterns from
existing tunes. Some of these tunes sound good enough to
convince audiences that they were composed by humans. As far
as we know, none of these AI-created tunes have broken into
the charts, but it is surely only a matter of time.

These AI tunes may be catchy. They might even be original,
in that their particular combination of notes and instruments
has never arisen before. But whether they have *soul* is another
matter altogether.

---

* 'Brown' here is short for 'Brownian motion, the jiggling motion of molecules as they are bumped
  by their neighbours.'
** There is no connection to the singer Pink, though her songs would be categorised as pink music
  (as would most pop music).

4

WHY WON'T THE CASE FIT IN THE BOOT?

How to squeeze things in or keep them apart

No matter how hard some of us try, when we have to fit the holiday luggage into the boot of the car there always seems to be one bag that is determined not to fit.

Normally a bit of repacking and reshuffling of the bags solves the problem, but fitting things into a tight space is a serious problem in the business world, and packing problems have been a source of investigation for many years.

One of the oldest such problems is that of packing circles into rectangles, a problem encountered by supermarket staff when stacking or packing tins of, say, baked beans. The tins have a circular cross-section, and have to be fitted into a rectangular shape. This may be the shelves they are to be displayed on, or the boxes they are transported in.

Fitting circles into squares

The obvious way to put circles together is in a rectangular grid, or 'lattice', like this:

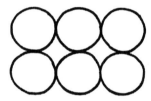

However, this is not the most efficient method. There is quite a bit of wasted space, and it's easy to work out what this wastage is. Remember that the area of a circle is π (about 3.14) times the radius squared. If the radius of a tin of beans is 5cm then its area is $\pi \times 25$, which works out at roughly 78.5cm^2. The area of the square around it is 10 x 10, or 100cm^2, so the circle is occupying only 78.5 per cent of the area.

There is a far better way of packing baked beans together, and this is the hexagonal method:

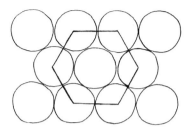

Here, the proportion of the area covered by the circles is just over 90 per cent. The exact figure is $\pi/(2\sqrt{3})$.

In fact, if you are packing tins of baked beans into a vast space, there is a mathematical proof, *Thue's Theorem,* which says this form of hexagonal packing is the densest packing possible. So, as is so often the case in mathematical problems, the optimal solution to a problem turns out to be based on one of the simplest known patterns.

However, Thue's Theorem is strictly true only if there is an infinite amount of space. The available space in the real world is most definitely finite. When space is limited, the most 'regular' packing is not always the most efficient. Take for example the problem of fitting nine tins of baked beans into a square tray. If you pack the tins in the normally optimal hexagonal layout, the smallest possible square tray that will hold them has to be about 3½ tin diameters across.

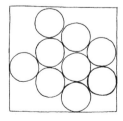

Compare this with the usually less efficient square array:

Here, the tins fit into a 3 x 3 square. In fact, in 1964 this was proved to be the optimal fit of nine circles into a square. You can't fit nine tins into a smaller square, however hard you try. Not unless you crush them, anyway.

As the number of tins increases, the optimal pattern changes. Some are quite surprising. For example, the optimal packing for thirteen tins is a square of roughly 3.7 x 3.7, in this formation:

It may look a bit of a mess, but it has been proved to be the best solution. Twelve of the tins are locked into position, while the thirteenth, the black one, is rattling around in the middle.

In everyday situations, it isn't that common to fit a box around the items being packed. Usually the box size is fixed, and the problem is to squeeze in as many items as possible. So mathematicians have also investigated the problem of fitting as many circles as possible into a square of some given dimension. This is subtly different from the problem that has been discussed up to now.

To make the dimensions easier to handle, let's switch from baked-bean tins to pennies. If the diameter of a penny is 1cm, how many pennies can you fit into a 10cm x 10cm square? You might guess that you can fit 100 pennies in, which is true if you use the regular square array.

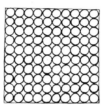

However, it is possible to fit in more than this. If you arrange the pennies in a hexagonal array, it turns out that you can squeeze in an extra five pennies, making 105 in total.

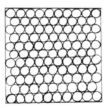

But even this isn't the maximum. By combining square and hexagonal packing, it is actually possible to fit another penny into the square. As with the earlier example, the optimum solution is not the most ordered and regular one:

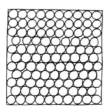

This type of packing problem is relevant not just for shops and packagers wanting to squeeze as many items into limited space as possible, but also for manufacturers who want to cut as many pieces out of a sheet as possible. One nice example of this comes from the shoe-making industry. In some upmarket manufacturers, shoemakers cut the leather uppers from cowhides by hand. Each cutter is given a target number of uppers to cut from the hide, and a cutter who exceeds the target gets a bonus. Perhaps you have replicated this problem in the kitchen with a pastry cutter.

So far all of the discussion has been about packing circles into squares. Just as common is the problem of packing rectangles into rectangles. Here, the obvious way to stack is to put the flat surfaces against each other, forming patterns like a chessboard, or the herringbone effect to be seen on many wooden floors.

However, the mathematician Paul Erdös discovered that such rigid and neat patterns are not always optimal. In exactly the same way that a bit of irregularity can help you to squeeze in more circles, sometimes more squares can be packed into a given space by twisting them around a little. Erdös produced a formula that said that in a square with side S cm and tiles with sides of 1cm, there is guaranteed to be a packing that leaves no more than $S^{0.634}$cm^2 uncovered. To put this in perspective, tiling a square of 100.5 x 100.5cm (area 10,100.25cm^2) with 1cm square tiles in regular chessboard format will leave a space of about 100cm^2 uncovered. However, by jiggling them in a less regular pattern it should be possible to squeeze in at least 81 more, leaving no more than $100.5^{0.634}$, or 18.6cm^2, uncovered.

The challenge of fitting squares into circles also has a practical application. The square silicon chips used for computers are cut from circular wafers of silicon. You can probably picture how the curved edges of the wafer end up as waste. However, the larger the wafer, the smaller the proportion of wafer that needs to be thrown away, which is why manufacturers have invested a lot of money in finding ways of growing larger silicon crystals.

Three dimensions – stacking a pile of oranges

When it comes to three dimensions, the problem of packing takes on a whole new level of complexity. The most analysed three-dimensional problem is that of stacking oranges.

Go to a fruit stall, and you might well see oranges being stacked in a pyramid like this:

Viewed from above, the oranges form hexagonal arrays layered on top of one another. In 1690, the great astronomer Kepler speculated that this was probably the most efficient way of stacking spherical objects (that is, the amount of air trapped between the oranges is the minimum possible). Nobody was able to find a better way of stacking spheres, but it wasn't until 1996 that it was finally proved to be true. Stacked this way, the oranges occupy $\pi/\sqrt{18}$ of the available volume (a little over 74 per cent).

This may make you wonder what would happen to the oranges if you were able to compress this stack together so that all of the air gaps were removed. What is the solid shape that each orange would be squashed into? Would it involve hexagons, like those produced when you squash together the circles on page 39? You can try to find out by squeezing together soft spherical objects, such as peas, freezing them and dissecting the result. (We tried it with marshmallows but they are much too elastic and bounce back to their original shape when unsquashed.)

It turns out that the squeezed oranges form into a peculiar regular solid known as a *rhombic dodecahedron*. This solid has twelve faces, each of them the shape of a diamond, with the following dimensions:

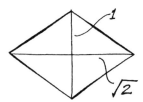

Viewed from some directions the outline of the rhombic dodecahedron is a hexagon, and viewed from others it makes a square. It is really quite a beautiful object.

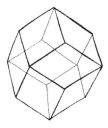

Because of its efficient packing structure, shapes approaching the rhombic dodecahedron can be found in some crystals, as well as in beehives and, presumably, in piles of soft, rotting tomatoes.

Packing into several boxes

Conveniently, everything we have been packing so far has been the same size. Again, real life tends not to be so simple. Whether you are filling the fridge with food or packing parcels into suitcases, the fact is that most of the time the objects to be packed are all shapes and sizes. In these cases, mathematicians usually throw their hands up in despair and leave it to operational researchers, who are less purist in their approach and are often content to come up with solutions that are reasonable but not perfect. The trick of the operational researcher is very often to develop sets of rules or *algorithms* that are guaranteed to lead to a solution that is within a certain target – say 10 per cent – of the best possible.

Think of those occasions where you are moving house. You know that the volume of items you need to shift is enough to fill about twenty identical boxes – except that it would require far too much time and planning to work out what to put where. Instead, you go for the easiest method of packing, which is to put items at random into the first box, and, when the next item you pick up doesn't fit, you seal up the box you have been filling and start on a new one. This is known as a 'first-fit strategy'. How efficient is it? It turns out that, however unlucky you are with the order in which things come to hand, the number of boxes you need will always be within 70 per cent of what could be achieved with perfect allocation. So if your best possible is 20 boxes, you can reassure yourself that even the lazy first-fit method strategy should require at most 34 boxes. If this isn't good enough – and, let's be honest, 70 per cent is a bit of a waste, though this is the *worst-case* scenario – a strategy of packing the biggest things first and the smallest last always turns out to be within 22 per cent of the very best solution. This means a worst case of 25 boxes, instead of the optimal 20. And, since many of us tend to use a biggest-first strategy, especially when filling the car boot, this shows that when it comes to packing, common sense is a good substitute for deep mathematical thinking.

Incidentally, this result doesn't just apply to putting objects into boxes. The things being 'packed' can be all sorts of other resources, such as money or time. Think of the people turning up at an embassy to have their visas processed. Because the circumstances of each case are different, each will take a different amount of time to be dealt with. These different service times are equivalent to the differently sized parcels. The officer behind the desk only has a certain amount of time at work each day, and that available time is equivalent to the box.

One rather bureaucratic embassy we came across dealt with people on a first-come-first-served basis, just like the basic box-filling method described earlier. If a new arrival at the desk was deemed to require more time than the administrator had remaining before clocking off for the day, the applicant was told to come back the next morning (i.e. today's box was full), and the desk closed for the night. This certainly seemed inefficient, though we can say with confidence that at most 70 per cent too many days were used to handle applications.

The attraction and repulsion of packing people together

What is great about most inanimate objects is that they don't mind being packed together as tightly as possible. This is not usually the case for people. The phenomenon known by psychologists as 'personal space' has all sorts of implications for the way people can be packed together. Another fundamental difference between people and baked beans, of course, is that the people usually have some influence on how they arrange themselves and so design their own packing rules on the hoof.

One of many classic people-packing problems is sitting down at the cinema. Unlike baked-bean cans or luggage, the audience arrives in dribs and drabs, and although the packing arrangement for a full house is predetermined, the way people occupy a cinema when it is not filled to capacity is a rather more complex problem.

Two forces are at work in determining where people sit in the cinema:

- *Attractive forces.* These are forces that are pulling people to sit in certain positions, though the strength of each force will depend on the individual. Some will be attracted to a position near the screen (so they can see), others to the back of the cinema (so they can engage in unobserved activities), and others are simply drawn to the most accessible seats (the end of the row, for example). This will lead to clusters of people building up sporadically around the theatre.
- *Repulsive forces.* By far the strongest repulsive force for people in cinemas is other people. This is the personal-space factor. Given a choice, people will choose to sit as far from people they don't know as possible, and will also avoid being directly behind somebody, so that their view is not impeded.

If the repulsive force were the only influence on cinema sitting, people would tend to seat themselves in something approximating a hexagonal lattice, which has cropped up before in this chapter, since that is the spacing in two dimensions that puts people as far from each other as possible. However, because of the various attractive forces, the hexagonal spread will be distorted, with denser packing near the front, back and central aisle of the theatre.

A similar combination of attraction and repulsion occurs on a bus. Here the attractive forces are primarily convenience.

Seats nearest the door or the top of the stairs are preferable for most people, though the front seats on the top deck have a special charm for some of us. The repulsive forces on a bus are stronger than those in a cinema, so the spread of passengers tends to be more even. Informal observation would suggest that single passengers always choose empty pairs of seats if they are available.

When all the customers or passengers are seated on the bus or in the cinema, the pattern they form could be called the *equilibrium* state. All this may sound vaguely like some of the stuff taught in physics lessons. There is certainly a tenuous analogy between the way people sit themselves and the forces between atomic particles.

How to get on to a tube train

A lot of research has been conducted into the dynamics of crowds and how they pack together and move. One interesting conclusion from this is that there is a best and worst place to stand if you want to get on to a tube train. Mathematical simulations have demonstrated that you will get to the doors far more quickly if you are on the platform edge moving along the side of the train than if you are front-on to the doors. The most obvious reason for this is that you tend to move more cautiously if you are at risk of bumping into people on either side of you. Moving along the platform edge, however, you have to worry only about bumping into the people on one side.

Total repulsion and the gentlemen's urinal

There are a few packing situations where the repulsion force is the dominant one. In these situations, the aim of the packing is for the objects to be spaced out from each other as far as possible. There is little point in having radio masts clustered together, for example, since their aim is to reach as wide a geographic range as possible. Since each radio mast broadcasts about the same distance in every direction, its reach can be represented as a circle. The aim of the network designers is therefore to cover the entire territory with as few circles as possible.

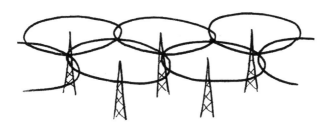

This spreading-apart problem is similar to the packing together of baked-bean tins, only this time no gaps between circles are permitted. Since every location must be within reach of at least one of the radio masts, circles will have to overlap. The optimal solution once again will be something close to the hexagonal lattice.

A more human spacing problem arises in the gentlemen's urinal. Women not used to the inner workings of this room may not be aware that the positioning of men at the wall is usually such that the spacing is as large as possible. If the urinal is empty, the first male to enter will tend to occupy a stall at the very end of the line. The next will occupy the stall at the opposite end. In order to keep the distances as large as possible, the third will therefore occupy the stall that is as near to the middle as possible. Each subsequent arrival is subconsciously (or maybe self-consciously) bisecting the largest space available. If the largest space is less than three stalls long, this inevitably means that a new arrival will be forced to stand immediately next to one or other of the current occupants. The repulsion force in this situation can be so strong for some males that they will divert to a cubicle instead.

In 2010, Evangelos Kranakis and Danny Krizanc wrote a paper entitled 'The Urinal Problem', in which they modelled the behaviour of men entering a urinal in order to work out the best place for the first male to stand if he wants to maximise his chances of maintaining his privacy when others start to occupy stalls. Their conclusion was that in most cases, the first user should go to the stall that is furthest from the door.

This behaviour is quite predictable. Whether the simple maths involved has any practical application outside the gents' toilet is another matter. Nor does it help to explain why women visit the ladies' in pairs.

5

SHOULD I PHONE A FRIEND?

How to make the big decisions in game shows

TV executives are always looking for new game shows to fill the schedules. Among the crucial ingredients for a hit show is the creation of some tension building up to a big climax. One way of building tension takes the form of an offer: 'Do you want to take the money, or gamble?' It is similar to the decision 'Take another card, or stick?' in many card games, including the vintage game show *Play Your Cards Right*. This is the $64,000 question, and, when it comes to those crunch moments in game shows, a bit of mathematics called *decision theory* can help.

Decision theory is used all over the place. It is particularly popular with government and management consultants. This chapter could just as easily have been devoted to deciding whether or not to build a new airport in London, or whether to invest in Asia or America, or countless other decisions. But for sheer human interest, you can't beat TV game shows for investigating the maths behind decisions.

Of all the many game shows, this chapter will concentrate mainly on two. In fact they are two of the most successful game formats of all time.

Can you afford to lose?

Imagine you are sitting in the chair of the quiz show *Who Wants To Be a Millionaire?* So far, you have done really well. You currently have £75,000. Your next question, if you get it right, will win you £150,000. If you get it wrong, however, you drop to the safety net that you've set – let's assume it's £50,000. Here is the question:

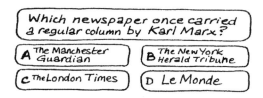

Not sure of the answer? You do have one lifeline left, which is '50-50'. This will reduce your choice to two and let's suppose you decide to take this lifeline. Now you are left with two choices. They are: (B) the *New York Herald Tribune,* and (C) *The London Times*.

Make your choice. You can offer an answer, or you can go away with the £75,000 you have already won. Only you can make this choice. And the answer? Find out later in the chapter.

Who Wants to be a Millionaire? stormed to international success in 1998. Like many successful quiz shows, *Millionaire* builds the tension by offering contestants the chance to gamble with important sums of money. The contestant has a choice of being greedy or playing safe. There is usually a hint of moral pressure on contestants to gamble, since this makes for much more exciting TV.

For those who have somehow escaped the mechanics of the prize money, here is how it is structured. Contestants start with nothing, and each time they answer a question correctly they move to the next prize money level. Here we'll use the levels that were used

£ 1 000 (first safety net)
£ 2 000
£ 5 000
£ 10 000
£ 20 000
£ 50 000 (second safety net)
£ 75 000
£ 150 000
£ 250 000
£ 500 000
£ 1 000 000

between 2007 and 2018 because they make the calculations we're going to do easier. The prizes from £1,000 upwards are shown above. Contestants who get a question wrong drop down to the safety net below their current level, so a wrong answer at, say, £10,000 means the contestant leaves with £1,000, while a wrong question at £500,000 means the contestant leaves with the second safety net, which (in the pre-2018 version of the show) was fixed at £50,000.

Decisions on whether to gamble or not are largely based on probability. You were just asked a question about Karl Marx, a man about whom you may or may not know very much. Your intuition will tell you how confident you are with the Karl Marx question, and you can convert that confidence into a probability. The simplest case is when you have no idea what the answer is. That way, you know that you are making a complete guess, and with a choice of two answers you have a 50 per cent chance of getting it right. The decision tree for a guess looks like this:

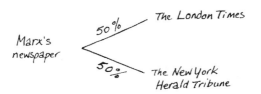

The branches of the tree represent the different outcomes, and along each branch is the chance of reaching that outcome.

> **Heads or tails – terminology quiz**
>
> *What is the chance of tossing a head on a coin?*
>
> *(a) ½*
> *(b) 50-50*
> *(c) 50%*
> *(d) 0.5*
>
> *The answer is … all four. Millionaire has popularised the expression 50-50 but probability experts use all four terms interchangeably.*

However, things become more complicated if you feel you know something about the answer. For example, you might reason: 'I know that Karl Marx lived for a time in London'. This may push you 75 per cent towards this answer, allowing a 25 per cent chance that you might be wrong. What this probability means, incidentally, is that in situations where you feel 75 per cent confident, you might expect to get the answer right three times out of four.

The decision tree would now look like this:

These probabilities are only gut feeling. There is no way of proving that the chance really is 75 per cent that *The London Times* is the right answer. After all, we know it is either 100 per cent right or 100 per cent wrong! In any case, your own value will be different, depending on what insights you have on the question. But in decision trees, you have to work with the information you have, and if all you can do is make an informed estimate then so be it.

Since you want to maximise your chances of getting the question right, you would in this example opt for the answer *The London Times,* because that is your hunch. But should you take the risk?

All you know so far are the relative chances of being right or wrong, but there is nothing to tell you whether to opt to have a go. To work this out, you need to put a value against the possible outcome of your decision. Decision trees can help you to do this. We already know the value of the outcomes of the Marx question:

- Get it right – £150,000
- Get it wrong – £50,000

Let's go back to the easiest case, where you really haven't got a clue which answer to go for. You might as well toss a coin to decide. You have a 50 per cent (or 0.5) chance of making the right choice. What does the decision tree tell you to do?

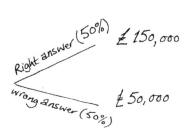

You will win £50,000 or £150,000* after this question, but on average your earnings will be somewhere in the middle. You can calculate what reward to expect by simply multiplying the money values on both branches of the tree by their probability. In this case (0.5 x £150,000) + (0.5 x £50,000) = £100,000.

What to do if you are a millionaire 'no-hoper'

The hardest part of Millionaire is actually getting to be a contestant in the first place. The odds are heavily stacked against you being invited to the studio, but, even when that happens, you are still competing with nine other participants to get to the Millionaire chair. And if you reckon the rest of the participants are far more knowledgeable than you, then you are in trouble.

But a bit of maths can help you. To get to the chair, you have to rank four items in order as quickly as possible. For example, place the following in order from most westerly to most easterly: (A) Paris, (B) London, (C) Norwich, (D) Brighton. Tricky, isn't it? Chances are that you are up against somebody who knows the right answer. Your only way to get to the chair is to get the question right, and submit your answer before everyone else. Trouble is, there are lots of possible permutations. It could be ABCD or ACDB, or any one of 22 other options.

If you think you are up against strong competition, your best tactic is actually just to bash the four letters in completely random order as quickly as possible, in less than two seconds if you can do it. That way, you can be confident that at least your answer was quickest Your chance of getting the right answer is 1 in 24 – about 4 per cent – but, if you miss out first time, there will almost certainly be a second chance, and possibly even a third during the show. In shows where three new contestants are chosen, your chance of getting the choice right and therefore being selected is in fact $1 - (23/24)^3$, or about 1 in 8. Not fantastic, but a darned sight better than the very remote odds you had when you started.

---

* Getting the right answer is actually worth more than £150,000, because the game doesn't finish there. If you reach the £150,000 level, you have the chance to play for £250,000 and beyond. For strong contestants, a better estimate of the financial value of getting the £150,000 question right is more like £200,000.

The value of playing the game is £100,000. To decide whether the game is worth playing, you have to compare this value with the alternative, which is keeping the money. If you do that, you go away with £75,000. Since £100,000 is more than £75,000, this simple decision model is telling you that if you have a 50-50 chance you should always gamble at the £75,000 level of the game. In fact, exactly the same maths says that it is worth gambling on 50-50 at every level in *Who Wants To Be a Millionaire?*, even when on £500,000 and going for the million.

But is this really what you would do? If you were offered half a million pounds, how certain would you have to be before you would risk losing it to go for £1 million? It all depends on whether you think £1 million is twice as valuable as £500,000. To most people, £1 million is more than they have ever dreamed of having. Then again, so is £500,000. The two prizes have very similar value or, to use the economist's term, *utility,* to most of us. Even £50,000 can seem a fortune to most people. To a very wealthy person, on the other hand, what's another £50,000? Its utility is relatively low. But another £1 million – now you're talking!

In other words, the value of the rewards in decision trees will be distorted when the prizes are big, and, unless you are looking purely at money for its own sake, you need to substitute utility values for money values.

To see how utilities can vary depending on the person, here are three stereotypical contestants:

- Angie: In debt, £8,000 would be life-changing
- Brian: Comfortably off, but £50,000 would pay off the mortgage
- Clarissa: Wealthy, but £1 million would deliver that longed-for yacht in the Bahamas

Maybe you can see yourself as being at one of these three levels.

Let's suppose utility is measured on a scale of 0 to 100. The utility graphs of the three might look like this:

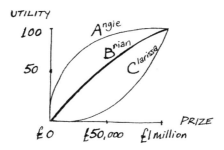

By the time that Angie gets above £50,000, the value of the prize begins to be almost irrelevant, because whatever it is it's enough to change her life. For Clarissa, on the other hand, prizes up to £50,000 are mere pocket money and worth almost nothing, but rapidly become more valuable as they move into six figures.

One model, rather too involved to reproduce here, that takes account of the different utilities at different levels of the game, comes up with the following suggestions on tactics, depending on whether you are type A, B or C.

The table shows how confident each contestant should be before they 'go for it' at each level:

PRIZE IF CORRECT	ANGIE	BRIAN	CLARISSA
£1000	80%	25% WORTH A 1 in 4 GUESS	25%
£10 000	90%	40%	25%
£50 000	95% TOO MUCH TO LOSE	40% GATEWAY TO BIG PRIZE MONEY	25%
£150 000	60% PAST CARING?	70%	50%
£1 million	90%	90%	75%

If every question in *Millionaire* meant you doubled your money or lost everything, then the players would have to be more and more confident as the stakes became higher to bother taking the risk. However, there is a safety net at £50,000. A contestant who gets a question wrong above that level is still guaranteed to take home £50,000. This question is therefore a crunch point for all of the players, and has a big distorting effect on players who are in a good position to gamble.

For Angie, every increment of money in the early stages is important, and so it would be foolish to throw away any prize unless she was very confident. Beyond £50,000, however, Angie can relax. The debts are all paid off. She can gamble a bit more.

For Brian and Clarissa, however, the big time doesn't really begin until the £50,000 level. Smaller prizes won't hugely affect their lives. It's even worth their gambling with odds of less than 50-50. Because the £50,000 level is a gateway to £75,000 and beyond, it's actually worth Brian gambling to get there even if his odds are worse than 50-50. And, because £20,000 is small fry to Clarissa, she gambles on the £50,000 question even if she is making a complete guess out of the four choices.

Curiously, according to this model anyway, it might be that by the time they reach £150,000, Brian is more cautious about risk than Angie. This is Brian's life-changing level, whereas Angie is well beyond that level, and will still be overjoyed to end up at £50,000. Clarissa will probably be a bigger risk-taker than either of them. Rich people will typically gamble more than those who are less well off at all levels in the game.

At the start of the chapter, you were posed the £150,000 question about Karl Marx. Did you keep the money (that probably makes you a Brian-type person) or did you gamble? The correct answer was: the *New York Herald Tribune* (a newspaper that had some left-wing leanings). If you got it wrong, you will be relieved this wasn't real money.

Are you the weakest link?

Millionaire's successor as the biggest talking point in the game-show world was *The Weakest Link*. Although the dynamics of this game are very different, in that contestants are competing with each other for a single prize, the mechanisms for winning money in the two shows have a lot in common.

The standard *Weakest Link* game starts with nine contestants, each of whom is asked a question in turn. If they get the question right, the prize money won by the team moves up one level. Before being asked a question, a contestant can also shout 'bank'. This puts the money won so far into a communal pot, and the value of the next question goes back to the start level.

The prize money of the original UK version of the show is relatively low. In fact the structure of the prize money in each round is as follows:

1	QUESTION RIGHT	£ 20
2	QUESTIONS RIGHT	£ 50
3	" "	£ 100
4	" "	£ 200
5	" "	£ 300
6	" "	£ 450
7	" "	£ 600
8	" "	£ 800
9	" "	£ 1000

Anyone watching this show must at some time have asked themselves: 'When is the best time to bank?' One theory is that when you get a question right, you should always take the next question and never bank, because the increments get bigger each time. On the other hand, if the group gets five questions right in a row and then the sixth one wrong, that is £300 thrown away. Maybe it is better to play safe and bank at low levels.

One way to analyse the game is to look at the prize money you expect to win by following different strategies. The analysis is complicated, because the decision trees become very involved. You can begin to see how you might figure it out, however, by looking at the expected winnings in the early stages of the game.

The most basic strategy is to bank money every time you get a question right, which earns £20 a time. Suppose you expect to get questions right 50 per cent of the time. Here is how your winnings would look after one round:

Question 1
    0.5 — *Answer 1*
         Correct £20
    0.5 — Incorrect £0

The average, or *expected*, winnings after one turn are (0.5 x £20) + (0.5 x £0)= £10.

What about after two questions?

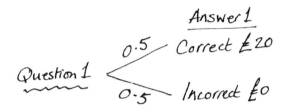

Answer 1   *Answer 2*   Amount banked

Question 1
  0.5 — Correct £20 — Question 2   0.5 — Correct £20   £40
                   0.5 — Incorrect £0   £20
  0.5 — Incorrect £0   0.5 — Correct £20   £20
                  0.5 — Incorrect £0   £0

There are four possible paths through the decision tree, correct/correct, correct/incorrect, incorrect/correct and incorrect /incorrect. To work out the value of this strategy, multiply the prize of each option by all of the probabilities along its branches, and then add together the values of each option. In each case, this is 0.5 x 0.5 x the banked money, which works out at an expected return of £20 after two rounds. In fact, with this strategy of always banking £20, and a 50 per cent chance of getting a question right, the expected winnings in the game will be £10 per question. After 25 questions, you would expect on average to have banked £250.

How does this compare with the next simplest strategy, which is to bank only when you reach £50? We are still assuming that you always have a 50 per cent chance of getting questions right. You might expect that, since £50 is more than double £20, this strategy will make more money.

After two questions, the possible outcomes are:

Always banking at £50 (50% answers correct)

Q1	Q2		Amount banked
CORRECT (0.5)	CORRECT	(0.5)	£50
	INCORRECT	(0.5)	£0
INCORRECT (0.5)	CORRECT	(0.5)	£20
	INCORRECT	(0.5)	£0

The expected return is (0.5 x 0.5 x £50) + (0.5 x 0.5 x £0) + (0.5 x 0.5 x £20) + (0.5 x 0.5 x £0) = **£17.50.**

Only £17.50? Remember, with the £20 banking strategy you expected to have £20 after two rounds. Surprisingly, after two questions, the £20 banking strategy is bringing in more money than the £50 banking strategy. And in fact, this continues to be the case no matter how many rounds you go through.

What happens if you expect to get, say, 70 per cent of the questions right. Does this affect which strategy is the best one? Yes, it does. Look at the decision trees after two rounds for the £20 and £50 strategies:

Always banking at £20 (70% answers correct)

Q1	Q2	Amount banked
CORRECT (0·7)	CORRECT (0·7) INCORRECT (0·3)	£40 £20
INCORRECT (0·3)	CORRECT (0·7) INCORRECT (0·3)	£20 £0

The expected winnings after two questions are (0.7 x 0.7 x £40) + (0.7 x 0.3 x £20) + (0.3 x 0.7 x £20) + (0.3 x 0.3 x £0) = **£28**

Always banking at £50 (70% answers correct)

Q1	Q2	Amount banked
CORRECT (0·7)	CORRECT (0·7) INCORRECT (0·3)	£50 £0
INCORRECT (0·3)	CORRECT (0·7) INCORRECT (0·3)	£20 (NOT YET BANKED) £0

The expected winnings after two questions are (0.7 x 0.7 x £50) + (0.7 x 0.3 x £0) + (0.3 x 0.7 x £20) + (0.3 x 0.3 x £0) = **£28.70**

So after two rounds, and with a 70 per cent chance of getting questions right, the £50 banking strategy emerges as marginally better. And this turns out to be true for longer runs of questions, too.

What also turns out to be true, however, is that, at this 70 per cent success level, always banking at £100 is even better than banking at £50. And banking at £200 is better still. In fact there is no success level at which banking at £50 or £100 is the best strategy, unless you are in the very final seconds, when you should always bank whatever you have.

Although the best banking level moves up in steps as the level of skill of the team goes up, crudely you can reduce the tactics for the basic form of *Weakest Link* to three very simple rules, as shown in the box overleaf.

Recommended group tactics for The Weakest Link

If you expect to get only half the questions right always BANK AT £20.

If you expect to get about two-thirds of the questions right BANK AT £200, and not before.

If you expect to get over 90% of questions right, aim for £1,000 and DO NOT BANK.

This is actually a practical strategy that you can use to 'play along' at home. You can usually assess a group as being a 50 per cent group, a two-thirds group or a 90 per cent group after just one round, and if you adopt the tactics used in the table when watching the programme, ignoring their *'Bank!'s* and using your own instead, you will usually end up with more money than they do. This is because weak groups push their luck instead of banking at £20, and because some strong groups tend to bank too early. In the final few rounds, even strong groups degenerate to the 50 per cent level.

If you were actually a contestant on the show, you would need to adapt your behaviour to reflect your personal circumstances. So, if you, personally, are 90 per cent confident of getting a question right, never bank. If, on the other hand, the people before you have built up to £450, but you only give yourself a 50 per cent chance of getting your next question right, bank immediately.

For higher levels of prize money, especially the US version of the show where the prizes are huge, the complications of 'utility' come into play just as they did in the analysis of *Millionaire*. And there are other psychological factors you can't ignore, too. For example, it's all very well taking a question at £450 without banking, but if you do happen to get it wrong then you are almost certain to be voted the Weakest Link, with all the humiliation that that entails. On top of this, there is the issue of not wanting to appear too clever. A contestant who excels in their answers is invariably voted off by the other two when the game is down to the last three players.

A three-card trick – and a tricky decision

There is one game show decision that has become famous not so much for the TV programme, but for the arguments about the right answer. It all centres on the finale of a popular American show from the 1960s called *Let's Make a Deal,* which was hosted by Monty Hall. At the end of the show, a contestant was shown three doors, and was asked to choose one of them. Behind one of the doors was a special prize, a car for example, and behind the others there was something of little value, a dustbin, say. Now imagine you are the contestant, and you choose one of the doors (Door 3). Before it is opened, the host opens one of the other two doors (Door 2) to reveal a dustbin. There are now two doors, and behind one of them is the car. You are asked if you would like to swap your choice of door. Do you stick with Door 3 or would you like to swap to Door 1?

Ninety-nine per cent of people when presented with this challenge stick with the door they first chose. The reasoning goes that it is a 50-50 choice, so why swap? But here is where it gets extremely tricky. It is not a 50-50 choice, because the host, being a showman, opened a door behind which *he knew there was a bin* in order to build up the tension. Your chance of picking the door with the car behind it was 1 in 3. The chance that it was behind one of the other two doors was 2 in 3. After the host has opened the door with the bin behind it, the chance that the other will have the car is still 2 in 3.

Now, experience says that the above short explanation is not nearly sufficient to convince the sceptics that door-swapping is a good idea. By far the best way to understand the problem is by experimenting. Here is one way to do it.

Ask a friend (let's call him Ralph) to deal three cards, including an ace, face down. Ralph should check so that he *knows which one is the ace*. Your challenge is to pick the ace, which represents the car. Choose a card, and then Ralph should turn over one card that he knows is *not* the ace, and then ask you if you want to swap or stick. OK, pick a card …

Here is the card you chose, still face down:

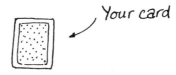

Ralph checks the other cards, then turns one over that he knows is not the Ace, like this:

He then asks you if you want to swap your card. If you decide to stick, you will win this game about one-third of the time. If you decide to swap you will win about two-thirds of the time, and certainly more than half the time unless you are very, very unlucky. Play this game at least ten times and you will begin to see why it isn't a 50-50 choice at all.*

Confusing? You bet! But this is yet another example where a bit of mathematics can make a big difference to your rewards. Then again, in the heat of a game show, which of us could truly keep our cool to figure it out?

---

* Although this is now almost universally known as the Monty Hall problem, it actually dates from the 1930s or even earlier. Indeed it seems very unlikely that the door-swapping situation described ever actually took place on Monty Hall's show. According to Monty Hall himself, 'On the show, I did indeed reveal what was behind one of the doors not chosen, but I do not recall giving the contestant an opportunity to trade her selected door for the one remaining. I asked members of my staff if they could recollect my ever doing so, and all but one of them said no.'

6

IS IT QUICKER TO TAKE THE STAIRS?

How to reduce waiting times for lifts

You might think that the biggest concern of a lift engineer is making sure that the lift doesn't break down. As it happens, though, designing a capsule that can dangle safely several hundred feet up is actually the easy part. The basic mechanics of a lift have hardly changed in fifty years, and, despite their reputation, modern lifts hardly ever go wrong.

Instead, the biggest issue with lift design is the time people spend waiting. The challenge has been devising ways of making sure that the lift picks up passengers and takes them where they want to go with the minimum of delay and frustration.

The problems that lift designers face are not dissimilar to those in other industries where waiting is an issue, such as supermarkets or traffic management. However, there is something particularly frustrating about waiting for a lift. With traffic and shops the problem is usually visible, and whether or not they are to blame there is at least somebody to shout at. This human element is missing in lifts. Behind those doors, a capsule is drifting up and down with an electronic mind of its own. For this reason, lift customers are particularly sensitive to the time they waste hanging around.

The quality of service is largely measured by the time interval between a customer calling a lift and the lift departing. Strictly speaking, there are two separate factors to consider. One is the *average* time that it takes a lift to arrive, and the other is the *maximum* time. Both have to be acceptable for the service quality to be satisfactory.

A lift with service times like this might be acceptable ...

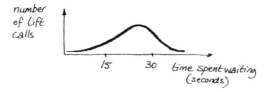

Whereas this lift service, with a lower average, might be unacceptable because of the occasional very long wait ...

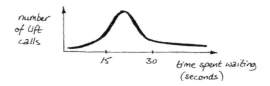

The first example has a high average but a small spread (or *standard-deviation*), while the second has a low average but a large spread.

> **How long should customers wait? Rule of thumb**
>
> *According to one lift company we spoke to, in a busy office environment lift users start to become impatient after about 20 seconds. After somewhere between 30 and 40 seconds many callers begin to regard the service as poor, and even the most patient users share this view after 45 seconds. Residential lift users are generally more tolerant than office users.*

The easy solution – build more lifts

The obvious way to keep the waiting time down is simply to provide lots of lifts. The more lifts there are, the greater the chance that there will be a lift on a nearby floor when a customer calls. However, having more lifts only helps if the lift cabins are distributed at different levels of the building. If all of the cabins are waiting on the ground floor, it doesn't matter how many lifts there are, as the following simple example demonstrates.

Suppose it takes five seconds for a lift to get to a customer, and customers are spread evenly throughout all the floors of a nine-storey building.

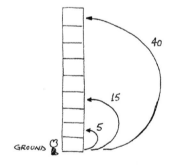

If all of the lifts are on the ground floor, the nearest lift will take zero seconds to reach a customer on the ground floor and forty seconds if they are on the top floor – an average of twenty seconds. This is the same if there is one lift or a hundred lifts.

However, with just three lifts spread at Floors 1,4 and 7, the diagram below shows that the average time it takes for a lift to reach a customer can be cut from twenty seconds to just over three seconds – a sixfold improvement in waiting time for a threefold increase in lift numbers.

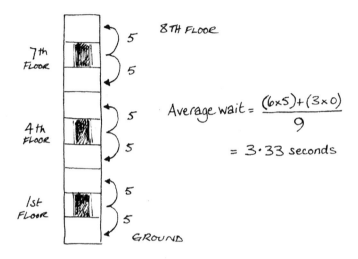

With nine lifts, of course, there can be a lift waiting on every floor reducing the wait time to zero.

Clearly, then, it is good for lifts to 'park' themselves through the building. There will be a tendency for this to happen naturally because of the random requests of customers, but lift designers prefer not to leave it to chance. To ensure that lifts don't cluster together, they are usually programmed to have a *zoning* system. Typically, each lift is allocated a particular range of floors that are its home territory, or zone. When not in use, a lift will return to its zone in much the same way as a dog returns to its basket. There it will remain with ears pricked, waiting for a call.

This brute-force approach of putting in lots of lifts and zoning them has two drawbacks. One is that lifts are expensive. The other is that lift shafts take up space. You can probably imagine circumstances where a company wants to house a thousand people in a building and to give them a rapid service by providing plenty of lifts – and then discovers that this can only be achieved if all of the office floor space is occupied by lift shafts! The shortage of space forces the lift designer to think more cannily, with the aim of achieving the best quality of service with as few lifts as possible. One solution adopted in some very tall buildings is to use double-decker lifts. In a double-decker lift, you can board either on Level 0 or Level 1. The Level 0 lift serves all the even-numbered floors, while the Level 1 lift goes to the odd-numbered floors. Because the two lifts are attached to each other, you only need one lift shaft, though there will be some inefficiency because, when the lift stops at Floor 40, those in the 'odd' cabin above have to stop at Floor 41 even if nobody wants that floor.

How many floors per lift?

Architects generally have a sense of how many lifts a building is going to need, but this will vary hugely depending on the type of building, and also on the ratio of its floor area to its height. A tall, thin office building might only need one lift for every four floors, but an office with a bigger footprint might have one lift for every two floors. A hospital, which will often have fewer floors over a large area, might have three times as many lifts as it has floors. Rather than leaving it to chance, the designers will usually produce a mathematical model of the flow of people in the building to understand what the demand on lifts is going to be.

Make the lifts go faster

If you can't have more lifts, another way of improving the service is to make the lifts go faster. This can mean literally increasing the speed at which the capsule travels. In some very tall buildings, lifts accelerate to a top speed of around 10 metres per second, or about 22 m.p.h. There is, however, an upper limit to how quickly they can reach this speed. Most customers don't take too kindly to big g-forces or sensations of weightlessness. (If they'd wanted that, they would have signed up for the space shuttle.) The rate at which a lift accelerates is therefore normally limited to about one metre per second every second, which means it takes at least ten seconds to reach a top speed of ten metres per second. There's a formula for working out distance travelled (first deduced by Galileo in the 17th century), which says:

$$Distance = \frac{1}{2} \, acceleration \times time^2$$

So in the ten seconds it takes to accelerate to full speed, the lift will travel ½ x 1 x 10² = 50 metres

Fifty metres represents about fifteen floors of the building. The lift will need another fifteen floors to slow down, which means that the building needs thirty floors just to enable the lift to reach top speed for a few moments. And that's assuming that the lift gets a clear run. In a busy building, a lift is likely to make lots of short trips, which means it will rarely have a chance to accelerate to high speed. Added to this, the time saved in speeding up the lift will be small compared with the time spent opening and closing the doors to let passengers in and out. In other words, making the lifts go faster has little impact on the overall waiting time.

The lift designer, therefore, has to find more subtle ways of speeding up the passenger delivery. One such way is the use of the express lift. In much the same way as train services combine intercity expresses with local commuter trains, lift systems in tall buildings have a combination of 'short-stop' and long-distance' lifts. Express lifts can reduce the average journey times of customers quite significantly, and the faster that customers can be delivered to their destinations, the less time other customers will have to wait before they are served.

Again, a simple example using our nine-storey building helps to illustrate this. Suppose it is morning rush hour, so all the demand for lifts is at the ground floor, and lifts are then delivering people evenly among all of the other floors, before returning to the ground floor to collect their next batch. There are two lifts, and each of them calls at all floors. As in the first example, it takes five seconds to travel between floors, and added to this we'll say that it takes ten seconds to unload passengers at a floor, making fifteen seconds per floor on the way up. The down journey takes five seconds per floor, or forty seconds total. The total round trip for the lift is therefore 160 seconds.

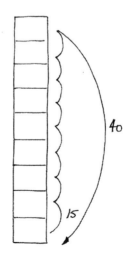

Now consider an alternative. Here, the first lift travels only to Floors 1, 2, 3 and 4. The second lift is now a fast service to Floor 5, and then calls at floors from 6 to 8.

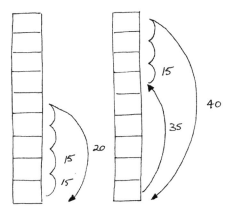

The cycle time for the first lift is 4 x 15 + 20 = 80 seconds. The second lift takes 5 x 5 + 10 + 3 x 15 + 40 = 120 seconds. In other words, *both* lifts now have shorter cycles than before.

Most tall buildings take advantage of this saving, with some or all of the lifts travelling to only a limited range of floors. In buildings of over fifty floors, there are usually additional lift lobbies, known as 'sky lobbies', too. Some of the lifts serve purely as expresses between the lobbies, at which point passengers switch to 'short-haul' lifts that serve their destination.

Anticipating the flow of traffic

Unfortunately, it's not enough just to provide extra-fast lifts. For the most efficient lift systems, the designers need to be able to predict the likely flow of people through the building so that the lifts can move to anticipate them. The pattern of human traffic will vary enormously depending on the function of the building and the time of day. In a standard hotel, for example, there will be a heavy flow of people between bedroom floors and the restaurant at breakfast time, and later there will be a steady flow from bedroom floors to reception.

The pattern within offices can be very different. Office blocks where different companies occupy different floors resemble hotels in that most journeys are to and from the ground floor. But, if a company occupies more than one floor, interfloor traffic is much heavier. The heaviest and most complex lift use is in buildings that are occupied by a single large employer – a hospital, for example, or a company headquarters – since the flow of people between floors could be just as heavy as the flow to and from the ground floor.

In the design of large buildings, it's common for the architects to make use of complex mathematical models to simulate how the traffic will behave at critical times of the day. The models will invariably include bits of probability theory to help them estimate how long a lift will take to deliver its load of passengers.

For example, suppose you enter a lift in the lobby of a building with ten floors above the ground, and five other passengers join you. How many times is the lift going to stop? If you are unlucky, you are heading for the top floor and every passenger decides to press a different button from yours. This means a tedious journey of six stops. Of course you might be extremely lucky, too. Everyone might choose the same floor as you, making this a rapid journey of just one stop. The average is somewhere in between, and as it happens there is a formula that enables you to calculate this.

If N people get into a lift at the lobby and the number of floors above them in the building is F, then the lift can be expected to stop:

$$F - F \cdot \left(\frac{(F-1)}{F} \right)^N \text{ times}$$

What does this mean? If there are ten floors and six people in the lift, the formula says that you can expect the lift to stop:

$$10 - 10 \cdot \left(\frac{9}{10} \right)^6$$
$$= 10 - 5 \cdot 31$$
$$= 4 \cdot 69 \text{ times}$$

In other words, for six people in a ten-floor building, there will typically be almost as many stops as there are people in the lift. But, as the number of people increases, the number of stops that you should expect on average doesn't increase so quickly. If ten people get into the same lift, the figures work out at:

$$10 - 10 \cdot \left(\frac{9}{10} \right)^{10}$$
$$= 10 - 3 \cdot 49$$
$$= 6 \cdot 51 \text{ times}$$

Four more people have got on, but the lift is only expected to stop one or two more times.

Incidentally, this formula of N people in a building of F floors is the same one that is used in a number of other analogous mathematical problems. For example, exactly the same formula predicts the number of different birthdays there will be in a group of N people. In this case F is always 365, the total number of birthdays available (ignoring leap years).

This 'F and N' formula is based on a number of assumptions, in particular that each floor or birthday is equally likely to be chosen. Since this is unlikely to be precisely true in the real world, the formula is only an approximation, albeit quite a reliable one.

Why lifts sometimes go the wrong way

A few old lifts operate using a relatively simple logic, and in these lifts it is occasionally possible to board, request a floor and then discover yourself travelling in the wrong direction. This is particularly the case in a type of lift known as the 'down-collective', which might be found in small hotels. These lifts usually have a single button outside, and they assume by default that the person calling them wants to travel downwards (as is usually the case in a hotel). If you are on floor five, say, and you call a lift as it is on its way from floor eight to a caller in the lobby, it will stop and pick you up 'en route', before continuing downwards. If you happen to request a higher floor, it will hold your request until it has completed its downward journey This can lead to the sort of farcical moment loved by comedians, where the scantily clad Romeo hoping to pay a visit to a lover upstairs boards an empty lift and finds himself embarrassingly carried down to the reception to be met by the contingent from the pensioners' Christmas party. Short of pressing the emergency stop button there is nothing he can do about it.

Lift logic

The logic used to drive lifts has become increasingly sophisticated, not only to reduce the waiting time but also to prevent some of that peculiar lift behaviour where the machine seems to have a mind of its own.

This behaviour can include, for example, lifts that appear to go in the opposite direction to the one that the customer requested (see the box opposite). Just as frustrating is the lift that seems to ignore the customer and just goes sailing by. In the early types of lift, this problem was caused because the lift was not capable of dealing with more than one instruction at a time. Until it had delivered its load, it ignored any other calls.

In more modern lifts, the most likely reason why a lift seems to ignore a caller is that the customer has asked to go down, and the passing lift is currently on an up journey. It could, however, be the case that the lift is full. Most modern lifts have weight sensors, and a fully loaded lift will not stop to pick up any more passengers, just as a full bus will zoom past a crowded bus stop.

The main difference between lifts and buses, of course, is that passengers can at least see that the bus is full. Lifts don't usually have an indicator to convey this information.

Lifts do, however, beat buses when it comes to advanced logic. For example, while a bus doesn't know how many people are waiting, in some office lift systems everyone waiting for the lift presses the button for the floor they want to go to so the lift knows in advance how many people are waiting and where they want to travel to.

However, even the highly sophisticated logic of the most up-to-date lifts can cause different types of behaviour that might appear illogical to the observer.

For example, imagine you are three floors down in the basement of a building with two lifts. The indicator tells you that there are lifts sitting on higher floors, one on the ground floor and one on the third floor. You call the lift, and notice that the one that comes to collect you is the one from the third floor – despite being twice as far away. Why didn't the ground-floor lift come to you? The answer is that 'intelligent' lifts are often

---

The lateral-thinking approach

There has been an underlying assumption in this chapter that the waiting time for lifts needs to be reduced. However, this is true only because people become frustrated when waiting. If people didn't get bored while waiting, then they would be less concerned about how long the lift took to arrive. According to office legend, one company with slow lifts got around the problem by putting mirrors outside the lifts. This didn't alter the speed of service, but customers spent the waiting time combing their hair and otherwise grooming themselves. The level of customer satisfaction rocketed. If this story is true, then the building manager deserves a medal for saving a fortune in lift engineering fees.

---

programmed to have a slight bias towards sticking to the ground floor, where most of the passengers get on. The intelligent lift may calculate that it's worth sending a more remote lift to collect you if it means that it can keep a lift waiting at the ground floor, where a flurry of passengers could arrive at any moment. You are being sacrificed (modestly) for the greater good.

Here is another possible sacrifice. An intelligent lift is seeking to keep down both the average and the maximum waiting time. A customer on Floor 6 calls a lift, but is dismayed when it bypasses them to collect somebody at Floor 9. The reason might be that the modern intelligent lift is aware that the Floor 9 person has already been waiting for a minute and is therefore top priority. With this urgent case on Floor 9, the lift reckons that you on Floor 6 can wait a few more seconds.

In other words, even the most complex logical system will at times fail to live up to the sometimes irrational and impulsive urges of us humans, and the faster the lift service becomes, the more we seem to demand from it. If it hasn't surfaced already, it won't be long before a 'lift rage' story hits the headlines.

7

HOW LONG IS A PIECE OF STRING?

The curious world of fractals

Here are two pieces of string. Which of them is the longer?

You are right, this IS a trick question.

The answer is B. These are not ordinary pieces of string. String A is perfectly straight, but look at B through a magnifying glass and it looks like this:

B is actually made up of tiny zig-zags, which make it twice as long as it first looked. But that isn't the end of it. Zoom in on any zig or zag and it looks like this:

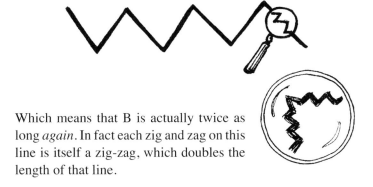

Which means that B is actually twice as long *again*. In fact each zig and zag on this line is itself a zig-zag, which doubles the length of that line.

And it turns out that this goes on for ever. Which means that, depending on how far you zoom in, and what ruler you use, the length of this line can be made to keep on doubling to anything up to infinite length. How long is a piece of string? It would appear that it really can be any length that you wish.

The infinite piece of string may be theoretical fun, but it actually has practical consequences in the real world.

Early in the twentieth century, it was discovered that the Portuguese and the Spanish were making different claims about the length of their border. This wasn't because of a dispute between the countries. Both countries happily agreed on the tortuous path that the border took, meandering for the most part along river valleys. But the measurements that they quoted in their reference books differed enormously. According to the Portuguese, the length of the border was 1,214 kilometres, while Spain claimed that it was 987 kilometres.

Discrepancies like this are commonplace. What is the length of the River Danube? Depending on which reference you use, it might be 2,706km (an edition of *Pears Cyclopedia* on one of our shelves), or 2,850km (Wikipedia). Meanwhile, according to the plaque at the river's official source in Bregtal, its length is 2,888km.

Why do the measurements turn out to be so variable? The answer is that the length of a river, or a coastline for that matter, depends on the accuracy of the map that you use. The more that you magnify the line, the more coves and bobbles you reveal, and within each bend there are other even smaller bends. A highly detailed surveyor's map will show up far more of these bobbles than, say, a road map. This can lead to huge discrepancies in the length of a line, as the zig-zagged piece of string showed.

Micro-patterns and fractals

There is something else important about the shape of a meandering river. Its winding shape looks much the same on a map of scale 1:10,000 as it does on a 1:100 map. This is similar to the zig-zag string, where zooming in revealed identical zig-zag shapes no matter how greatly the line was magnified.

How long is a piece of string?

The normal way to measure a piece of string is to pull it straight and measure the distance between the two ends. However, to measure its true length you need to run a ruler along its edge, and since its edge is not a perfect straight line even when it is pulled tight, the answer you get will depend on the accuracy of the ruler that you use, in just the same way as the length of a river depends on the scale of the map. So one answer to the question 'How long is a piece of string?' is 'It depends on the ruler that you use.' Other more facetious answers also exist, of which the most common is 'Twice the distance from the middle to the end'.

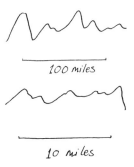

100 miles

10 miles

Patterns that continue to reveal similar patterns on a smaller and smaller scale as you zoom in have a special name. They are known as *fractals*. Fractals are not dissimilar to a Russian doll, where you open up the giant doll to reveal a smaller replica, inside which is another smaller but otherwise identical doll … and so on.

Shapes that are fractal, or certainly close to fractal, appear throughout nature. One often-quoted example is a fern leaf such as bracken. The large leaf is composed of nearly identical copies of itself:

Broccoli is another example. Take a large head of broccoli and you will find that it is composed of several branches. Cut off one of these branches, and the result is a smaller but otherwise perfectly formed head of broccoli. This too will be composed of branches, each of which leads to a mini-broccoli. You can often go to four stages with this, ending up with lots of cute little baby broccolis.

Shapes of this complexity can be produced from quite simple rules. Here is an example of the rules that you could use to make a treelike shape:

Start with a single vertical line (a branch), of length L: The rule for adding branches is as follows:

⅓ along the branch, add branches length ½ L, 30° on each side.
⅔ along the branch, add a branch length ⅓ L, 30° on each side

After one iteration, it looks like this:

Now apply the same set of rules to each new branch, and then to those branches. The result bears a close resemblance to a tree.

This example shows how fractal shapes can be produced by following a sequence of simple, repetitive rules. However, the mystery of fractals deepens when it turns out that fractal shapes can sometimes be produced by apparently random sequences, too.

Here is an odd little game. Below is a triangle, with corners labelled A, B, C. The object of the game is to fill the triangle with dots.

Start by choosing a point anywhere at random inside the triangle. An example is shown at the point X, though you can choose any point you like.

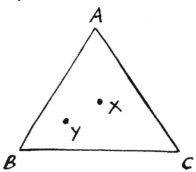

To determine where you place your second dot, you need to choose randomly between the three corners A, B and C. One way to do this is by rolling a die. Numbers 1 and 2 represent A, 3 and 4 B, and 5 and 6 C. Suppose you choose B. The next point must be drawn exactly midway between your current position and B. In the triangle on the previous page, Y is the second point. Roll the die again to determine the point to follow Y.

Keep repeating this process for as long as you can, trying to be accurate with your plotting. After twenty or thirty goes, a pattern begins to emerge from the dots you have drawn, and, the longer you go on, the clearer this image becomes. Surprisingly, the pattern is not completely random. Instead, it is a series of intricately nested triangles that get smaller and smaller. This pattern is known as the *Sierpinski gasket,* and the shape is fractal. No matter how far you zoom in, you will find replicas of the pattern of inverted triangles.

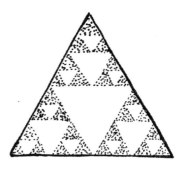

The pattern was created by an apparently random process of plotting dots. However, this same pattern can be produced in a completely different way that is not random at all. Draw an equilateral triangle, and shade in an upside-down triangle in its centre, like this:

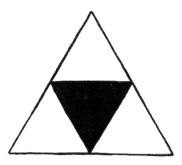

Now do the same thing in the three remaining white triangles. Repeat this process on all of the smaller white triangles that you create, and the result will be the gasket once more.

This connection between ordered rules and apparent randomness is an important feature of many fractals, and doesn't just apply to shapes. It can apply to numbers, too.

Fractals in number patterns

What can it mean for number patterns to have fractals within them? All the examples of fractals so far, from rivers to broccoli, have been demonstrated in pictures, and the best way of seeing fractals in number patterns is to represent these as pictures, too.

The most common way of representing numbers in a picture is on a graph.

If you want to know the practical applications of fractal mathematics, you should ignore what follows and jump straight to the next section. However, it is interesting to divert from everyday life for a moment to see how algebra can be intimately linked with fractals.

Before creating a fractal image, we need to generate the numbers that will go into it. There are many ways of using mathematical formulae to produce fractal shapes, and some of these are extremely complicated. What follows is one of the simplest, and you may find even this a little bit fiddly, but it's worth working through it just to appreciate the surprising pattern in the end product.

To produce the fractal, we're going to need to generate numbers by feeding them into a box, then take what comes out at the far end, and feed that back into the box. This is known as an *iterative function*.

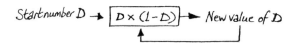

Choose any decimal number D between 0 and 1, and feed it into the box. For example, the starting point for D might be 0.6. The rule for producing the new value for D is to take the current value of D and multiply it by $(1 - D)$. In this case, $0.6 \times 0.4 = 0.24$.

Now feed this new value of D, 0.24, back into the box, to create the next value of D: $0.24 \times (1 - 0.24) = 0.1824$. Repeat this cycle a few times, and D will rapidly decay away towards zero. In fact this will happen whatever value of D you start with. Nothing exciting so far.

Things get much more interesting, however, if another box, K, is added to the system.

K is like a control button on the system, and we're going to see what happens to the final value of D as K is increased.

We already know that if K is 1, D always decays to zero. Suppose instead that K is 2, and you choose a starting value of D as 0.6 like last time.

First time through the system:
$$0.6 \times 0.4 \times 2 = 0.48 \dots$$
… feed this back through to give …
$$\dots 0.48 \times 0.52 \times 2 = 0.4992 \dots$$
… feed this back to give
$$\dots 0.4992 \times 0.5008 \times 2 = 0.49999 \dots$$

After just three loops through the system, it's already becoming apparent that D is going to end up at 0.5. What's more remarkable is that, if K is 2, whatever value of D you start with between 0 and 1, it will always end up at 0.5, and it will get there pretty quickly, too. Before going any further, you might care to try this out for yourself by choosing a different starting number for D. A calculator would be handy.

What will happen if K is 2.5? D always ends up at 0.6, whatever value of D you start with.

If K is 3, however, something odd happens. D eventually ends up oscillating between two numbers, 0.669 and 0.664. *Oscillating?* How on earth has this happened? It's very mysterious, and this is only the start of it.

When K is 3.47, D always ends up cycling around four numbers, 0.835, 0.479, 0.866 and 0.403. From here, as K increases by small increments the quantity of numbers in D's final cycle doubles with rapidly increasing frequency. First D oscillates between eight numbers, then 16,32 and so on. This process of doubling is called *bifurcation*. Finally, when K gets close to 4, there seems to be no cycle at all. D just leaps in a seemingly random fashion from one number to another, never settling anywhere.

Value of K	Final value of D
1	0
2	0·5
2·5	0·6
3	Oscillates between 0·669 and 0·664
3·47	Oscillates between 0·835, 0·479, 0·866 and 0·403
Close to 3·6	"We have total mayhem, captain"

Here, at last, are some numbers to plot on a chart, though to produce an accurate diagram you'd need to examine every value of K between 1 and 4. This is because between areas of apparent randomness, there will occasionally be tiny zones where the number of oscillations drops to a small number again – for example, at 3.74, there are five numbers in the oscillation.

This is an impression of what the complete diagram would look like:

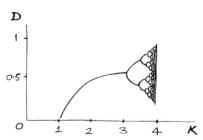

The lines represent the values that D ends up at. Notice how the single line suddenly splits into two lines when K reaches 3, and continues to split thereafter.

So where is the fractal? If you zoom in to look at any part of this chart in more detail, you will find that it is made up of micro-versions of the same complex pattern.

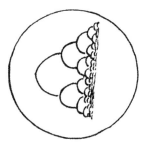

The simple act of multiplying D by 1 – D has created a fractal of quite remarkable intricacy. And we've seen that inside this ordered pattern there are patches that seem almost 'chaotic' – a word we'll meet again in Chapter 8.

This seemingly abstract example has a surprising link to real life: it can be used to model how animal populations fluctuate.

In the formula at the top of page 91, 'K' was the control button to see what happened to the output 'D'. The value of K can be used to represent the fertility of, say, rabbits, and D is the size of the rabbit population. If K is too small, the population dies out completely, but as fertility goes up, the rabbit population grows – until it reaches a level that is unsustainable. Too many rabbits means not enough food to go around so rabbit numbers plunge. But in the next generation the smaller population does have enough food, so the population grows. Our model predicts that there are certain fertility levels that can lead to population numbers moving up and down quite radically from year to year in a regular fashion – and this is what is seen in practice.

How fractals helped the Internet ...

Now that you know that mathematical formulae can produce fractals, there may be a little question nagging away: 'This is all very pretty, but so what?' In the 1990s, fractal geometry was put to valuable use in speeding up the transmission of images on the Internet.

If you are in the habit of downloading pictures from the Web, you will be well aware that this can be painfully slow. This is because the amount of information stored in a picture is huge, and transmitting details of every pixel can mean an image occupying hundreds of kilobytes. To cut down on this, programmers needed to get clever. We've already seen that mathematical equations or rules can create complex images. Is it possible that a picture of Buckingham Palace or Tom Cruise could be reduced to a formula, too? After all, a formula takes up far less space than the entire details of a picture. Without stretching the truth too far, this is just what has been achieved.

Look again at the image of the tree on page 87. The slow way of reproducing this picture is to copy each point in turn. A quicker way, however, would be to recognise that the big image is really just multiple copies of one of the smaller branches. To rebuild the image, all you need is one small section, and instructions on where to copy this in order to create the whole image.

Exactly the same principle applies with more complex pictures. All published images are really a combination of tiny coloured dots, most easily seen if you study a photograph in a newspaper. A crude image of Tom Cruise can be created using big coloured dots, and a fine image by using tiny dots. If you search for long enough, you will find that patterns in the crude image can be found in tiny sections of the fine image. For example, the crude image of Tom Cruise's nose could be exactly the same as a tiny portion in the fine image of his earlobe.

By searching through the image, it's possible to find close

matches between every large part and some corresponding tiny fragment. With the appropriate instructions (for example 'rotate by 45 degrees and shrink by a factor of 10'), and after several iterations, the crude picture can be used to rebuild the much higher-quality picture. In the end, this 'fractal image compression' lost out to other image formats, but the idea of looking for similarities is still used in video compression – most notably in MPEG, which is used for sending video over the Internet.

... and how fractals could earn another fortune

There is one other area where understanding the maths behind fractals could lead to even more of a fortune.

If you are a shareholder, you probably take an interest in the Dow Jones Index or the FTSE 100. Every day the index moves up and down, and every week, every year, and so on. Predicting the movement in the price of shares is a lucrative business if you get it right, so not surprisingly city analysts have spent millions upon millions trying to come up with ways of forecasting which way the prices will go. The trouble with most forecasts, however, is that they only seem to work after the event. A perfect model can be produced that exactly mimics what happened in the past, but try extrapolating it to predict the future and the results are often little better than sticking a pin into the page at random. (There have been numerous experiments to compare the performance of 'expert' stock market analysts with complete amateurs – including monkeys. In 1999 a six year old chimpanzee called Raven picked stocks by throwing darts at a list of 133 Internet companies. Her portfolio more than trebled in value, beating 6,000 professional Wall Street brokers. Similar results have been replicated many times since using random stock-picking techniques.)

The problem with stocks and shares is that, while their long-term trends seem to be reasonably steady, their short-term movements have always appeared random. But the growing interest in fractals has led analysts to take another look at the patterns of moving share prices. Like meandering rivers, the patterns of share prices show signs of being fractal.

Take a look at the price of a particular share over the course of a year:

Here is the same share price viewed over shorter time intervals during a week:

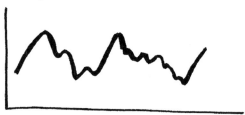

And this is what it looks like measured in the course of a single day:

The shapes are all very similar. It's as if we are zooming in on a fractal line.

However, observing a pattern is one thing, making use of it is quite another. Can the knowledge that share prices are fractal help you to forecast what will happen to a share in a year's time? If the graphs of share prices really are fractal, then maybe a formula for that graph can be extracted. Some analysts have suggested this might be possible, though we are not aware of any such formula being published. Of course, if there really is a way of using fractals to predict share prices, then the people who have found it will want to keep pretty quiet about it.

The fluctuation of share prices is just one of several occurrences of seeming randomness in this chapter. The meandering of a river and the oscillations produced by a simple number function were two others. This randomness associated with fractals has a special name. It is called *chaos*, and deserves a chapter of its own …

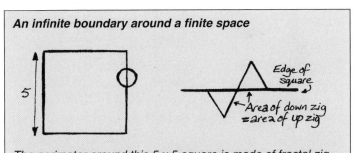

An infinite boundary around a finite space

The perimeter around this 5 x 5 square is made of fractal zig-zag lines of infinite length, like those at the start of the chapter. However, the area of the square is not infinite. Every bit of area lost to a 'down'zig is compensated for by an 'up'zig of the same size. So the area is 25cm². Here, then, is an apparent paradox: it is possible for a shape to have a finite area, even though its perimeter is infinitely long.

8

WHY DO WEATHER FORECASTERS GET IT WRONG?

Unpredictability and chaos

It has often been said that, while most countries have climates, Britain just has weather. If there was an index that measured the degree of fluctuation from rain to sunshine, windy to still, and warm to cold over short periods of time, the British Isles would surely top the league.

This is one reason why weather is the most popular topic of conversation, and why forecasts are a key feature of every news broadcast and achieve audience ratings unmatched in any other country.

You would think, though, that with so much weather about, forecasters would have got the hang of it by now. So why is it that weather forecasts can sometimes get it so horribly wrong?

The answer begins not in the sky but on a pool table.

Big breaks and flukes

You probably know the normal way to begin a game of pool. The white ball is placed at one end of the table, and is then

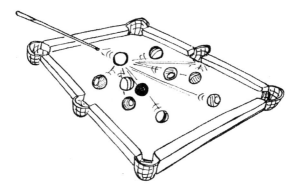

struck as hard as possible to smash into the triangle of striped and spotted balls at the other end of the table.

The balls scatter seemingly at random across the table, and if you're lucky one or two of them drop into the pockets.

Apart from the need to make a decent collision between the white and the colours, the opening shot is almost completely without skill. That is to say, there is no way of predicting with any confidence where all the balls will end up even if you attempt to hit the white with exactly the same strength and in the same direction each time. The tiniest change in the way you hit the ball or the way that the triangle of balls is set up leads to a different outcome.

It seems fair to describe the spreading out of the coloured balls at the start of a pool game as chaotic, and in fact *chaos* is exactly the term that mathematicians would use to describe it. Because it is a relatively new science, mathematicians are still developing new definitions of chaos, some of them extremely complex. However, there is one underlying theme to chaos that most mathematicians agree on. Something is chaotic if the tiniest of changes to the initial input can lead to a completely different, and unpredictable, outcome.

This effect of small errors having huge consequences has been known for a long time. Benjamin Franklin, one of the founding fathers of the USA, was responsible for this well-known quotation:

> *For the want of a nail, the shoe was lost;*
> *for the want of a shoe the horse was lost;*
> *and for the want of a horse the rider was lost,*
> *being overtaken and slain by the enemy,*
> *all for the want of care about a horseshoe nail.*

Might it be that some battles – and the whole course of history thereafter – have been influenced by the tiny detail of whether a nail was missing from a horse's shoe? Many would argue that they have.

A contest that is closer to most people's experience is a game between two teams. As in battles, everyone knows that the outcome of sporting tussles can hinge on one small incident, such as a player being booked, or a ball deflecting at just the wrong angle. However, what cannot be predicted is what would have happened had the hinge moment gone the other way. When commentators say 'the game finished one-nil, but it should have been *three-nil* because of those two missed chances' they are wrong. Suppose the first chance had gone in, instead of hitting the post. Although the score in the game at that point would have been 2-0, what would have happened thereafter, says chaos theory, would have been unpredictable. Yes, the team in front would have felt stronger at that point, but the different interactions and tactics thereafter could have led to a 5-0 win or a 2-2 draw, or just about any other result.

How is that?

Perhaps one of the best illustrations of how sport can show chaotic tendencies – that is, small changes in the starting conditions producing large, unpredictable changes in the outcome – comes from a computer program to simulate a cricket match. This program, written by Gordon Vince in the 1980s, allows you to enter the names and details of any two teams, and then to play a complete cricket match between them. In fact this program is really an extension of a dice game called Howzat, which was popular long before the computer era.

Every event in the match is simulated by functions with random elements – a computerised version of throwing dice for every incident. For example, from any ball a batsman might score runs, or he might be out, or nothing might happen (there's plenty of that in cricket). The program is extremely realistic – bowlers become 'tired' and perform worse when they have been bowling for a long time, and batsmen become 'nervous' and are more at risk of getting out when their personal score approaches 100. The program produces a complete printout of what happens after every ball of a match, and the outputs look convincingly like real matches.

To generate a match from this program, you need to enter details of the two teams, including a strength factor for each of the players. You also enter a seed number, which acts as a random-number generator. In effect, the seed decides what the outcome will be of every die-throw throughout the game. A different seed leads to a completely different match.

As an experiment to see how predictable the result would be, we used the program to simulate a cricket match between England and the West Indies. Details of the two teams were entered, with each batsman being given a strength factor, a number between 5 and 40. This strength factor determined

whether the batsman was likely to score a lot of runs or hardly any runs. Using a seed number of 444, the scores in the match looked like this:

West Indies 1st innings: 193
England 1st innings: 162
West Indies 2nd innings: 253
England 2nd innings: 187

A quick bit of adding up will confirm that the West Indies scored more runs overall than England in this match. In fact the West Indies won by 97 runs.

The match was then replayed using the same seed, and with all of the player details the same except for one. The 'strength factor' of one West Indies batsman was increased from its starting position of 23 to 25. Because this batsman's batting strength had been increased, the overall strength of the West Indies team was now slightly higher than before. Everything else was unchanged. As a result, you might expect the West Indies to have won the game and by an even larger margin. Yet the outcome of the new match was as follows:

West Indies 1st innings: 244
England 1st innings: 525
West Indies 2nd innings: 332
England 2nd innings: 52 for 0

Even though they were relatively stronger than before, the West Indies' performance actually got worse while England performed far better this time. In fact, to use the cricketing parlance, England won this match by ten wickets. This isn't an unusual example. Tweaking any of the starting factors, sometimes by very small amounts, could have had a similarly huge impact on the final outcome.

Why should this be? Suppose that because the batsman is stronger than before, he scores a run at one point that he would otherwise not have scored. This brings his partner to face the bowler, and since the partner has a different batting style, the next ball has a different outcome – maybe the batsman is bowled out on this ball, which would otherwise not have happened. The resulting chain reaction of events causes the game to diverge increasingly from its original path until it becomes unrecognisable from its previous incarnation.

This system is behaving in a chaotic manner. Whatever his knowledge of the game, no expert could have predicted this outcome. Instead he would end up with egg on his face, muttering the old adage used in so many sports: 'It's a funny old game'.

The pendulum and the magnet

Another quite different situation that illustrates how tiny variations in the starting point can lead to big changes in the outcome is a toy that used to be common on the desks of executives.

The pendulum-magnet is a mesmerising device. A ball hangs on a pendulum above a steel base. On the bottom of the ball is a magnet, and on the base are three more magnets, each of which is placed so that it will attract the ball towards it.

To start the motion, the ball is pushed aside and then released. Without the magnets, the ball would just swing to and fro, but because each magnet is pulling it, the ball instead swings all over the place, sometimes in a violent jerking motion, until it eventually comes to rest above one of the three magnets, which we'll call A, B and C for old times' sake.

The three base magnets can be arranged in a triangle in such a way that each of them ends up with the ball above it about one third of the time. However, just like the pool balls at the start of the chapter, it can prove very difficult to release the ball and predict with any confidence which magnet it will end next to in

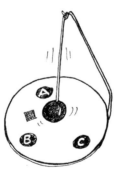

any particular turn. One time it will end above magnet A, but the next time, despite being released from apparently the same place, it ends instead above magnet B.

An insight into why this unpredictability arises came when mathematicians were able to produce a computer simulation of the pendulum. Since all the dimensions and forces involved were known, it was, as these things go, a reasonably straightforward

Randomness: how can a computer simulate dice?

Most games programs on computers require the computer to do 'random' things. Any computer, therefore, needs to be able to produce on command numbers that are as unpredictable as the outcome of rolling dice. This is not as easy as it may sound, because the whole point of computers, of course, is that they are there to be predictable by following rules.

Although they are not capable of generating truly random numbers, all computers contain a formula that will produce 'pseudo-random' numbers – numbers that appear random even though they are generated by a precise sequence of calculations. Many methods exist for generating such sequences, (the formula on page 91, with K = 3.8, is one crude example). Usually these methods require an initial seed number to start them off. This seed can either be entered by the user, or else taken from the computer's clock (for example the number of seconds past the minute at the instant when the keyboard is struck).

Randomness is quite tricky to define, but a common way of testing for it is to check that (a) all numbers in the sequence appear roughly the same number of times as each other, and (b) the numbers do not follow any predictable pattern. The sequence 1, 2, 3, 4, 5, 6, 7, 8, 9, 0 passes the first test of randomness, but fails the second. 5, 8, 3, 1, 4, 5, 9, 4, 3, 7, 0 appears, at first glance anyway, to pass both tests, though it is only pseudo-random because it was in fact generated using a simple rule -can you work out the rule?

task to write a computer model mapping out the path of the ball right the way through to its finishing point. That way, the end point of the ball could be calculated for any specific starting point.

The results were surprising. Suppose you map out some of the possible starting positions of the ball in a square grid. If the ball is released from the midpoint of each square on the grid you can work out which of the three magnets it will end above. Here is how part of the grid might look:

If the ball is released from anywhere in the zone on the left, it seems certain to end up at B, while the upper middle area or the right seems to be an 'A' zone.

However, if you zoom in on the B square that is highlighted, and divide it into smaller squares, then a whole new complexity is revealed within:

It turns out that, although the ball ends above B if you release it from the middle of this square, nearby squares can lead to other end positions. Nestling within what was a 'B' square is a cluster of Cs and a solitary A. And zooming into any of those squares reveals another pretty and yet unpredictable pattern made up of As, Bs and Cs. However far you zoom in, fresh patterns will emerge. No wonder that tiny errors in the starting position of the ball can mean that it ends up at a different magnet from the one that you predicted.

Note that not all of the zones will produce these chaotic results. There will be some solid regions where the endpoint will always, consistently, be A, B or C. These are, if you like, the predictable zones. Other areas, like the one described above, are chaotic zones.

The ever-changing patterns of As, Bs and Cs in the chaotic zones will have a familiar ring to them if you have read the previous chapter. In fact fractals and chaos are intimately linked together, which is why each of them has managed to infiltrate the other's chapter.

Chaos in the weather

Pool balls, missing nails and magnet-pendulums contain all the analogies we need in order to explain the problems in forecasting the weather. As with the magnet-pendulum toy, weather is created by a range of simple forces combining together. In the case of the weather, these forces are caused mainly by the sun's heat and the earth's rotation. Small changes in these forces can in some circumstances have enormous impacts on the weather.

In fact, one of the earliest discoveries of chaos came when a researcher called Edward Lorenz attempted to model the way in which weather patterns develop. Using a computer, he devised a relatively simple model to simulate how a weather system would evolve given certain starting conditions. The computer then generated reams of paper with numbers showing the changes in weather patterns.

When he wanted to rerun the simulation, Lorenz decided to save time by copying down the numbers halfway through the process and using those numbers as the starting values. To his amazement, however, the weather forecast changed, even though he had copied down the correct numbers to several decimal places.

It turned out that the numbers that the computer had printed out included rounding errors. What was printed as 17.427, for example, was really 17.42719163 in the computer's memory. That tiny error was enough to produce enormous changes to the outcome of the weather forecast for a week later. This was genuine chaos at work – a simple system creating highly complicated and unpredictable patterns. Thanks to his observations, Lorenz later made the claim that a tiny perturbation such as a butterfly flapping its wings could, in the right circumstances, lead to a chain of events that would create a hurricane in Florida – in just the way that a missing nail could have brought down George Washington's army.

Fortunately, even though a tiny change to the starting conditions can have unpredictable consequences, the overall weather patterns turn out normally to follow certain reasonably well-known routes. By running several different simulations, all using slightly different starting conditions, forecasters can see what the weather outcomes might be. Often, all of the outcomes will be similar to each other, indicating that the weather is in a predictable zone. Sometimes, however, small changes to the starting conditions lead to a wide range of weather predictions. Here, the weather has entered a chaotic zone. The longer in range the forecast, the more likely it is that a chaotic point will be reached, after which the forecast becomes little more than a guess. That's one reason why British forecasts rarely extend beyond five days.

And, whatever the forecasting model says, there is always a chance of a freak outcome. Michael Fish will for ever be remembered as the weatherman who publicly reassured a viewer in October 1987 that a hurricane was not due that night. Twenty-four hours later, southern England had been flattened by the worst storm in living memory.

A sports commentator would have known exactly what to say. It's a funny old climate.

9

WILL I CATCH FLU NEXT WINTER?

Epidemics and how they spread

Stories of the Black Death and the Great Plague still carry a morbid fascination for us hundreds of years after they happened. Both were probably the result of the same thing, a bacterial infection called *Yersinia pestis* carried by rat fleas. The plague was so infectious that it took only a few cases to start this horrendous epidemic.

In fact the Black Death entered Europe when a Tartar army catapulted infected corpses into a Genoese trading post, a scene of almost Monty Pythonesque absurdity were it not for the grim consequences. About a quarter of the European population died.

Plague is still out there, but fortunately its effect on mankind has been massively reduced in the last century, thanks to the advances in medicine and hygiene. Another important factor in the control of this and other diseases has been the science of *epidemiology,* which seeks to understand how epidemics work, and how they can be managed. Epidemiologists are in demand today more than ever. AIDS, swine flu and most recently the coronavirus have all been major world threats, and mathematical modelling has been an important weapon in fighting them.

The infectivity of gossip

It isn't only disease that spreads through communities. One of the most familiar and everyday epidemics is that of news and gossip. So, as an ideal introduction to the topic, let's take a look into the world of spreading news. Imagine that you hear a bit of hot scandal. Knowing that it would be unkind to tell too many people, you pass the news on to just a couple of your closest friends. 'Don't tell anybody,' you say. 'Certainly not,' they promise. However, they are naturally unable to keep it completely to themselves, so they, too, allow themselves to disclose the secret to a couple of confidants, under the strict agreement that they will keep it quiet. Those confidants do the same, and so it goes on, each new disclosure leading on to 'just a couple' of others. For the sake of argument, let's suppose that the news broke at 8 a.m., and that each person's disclosures took place within half an hour. How many people will know the news by 8 p.m. the same day?

8:00 a.m. Only you know the news
8:30   You and two friends know (1+2)
9:00   You, your friends, and their friends know
          (1 +2 + 4)
9:30   Eight more have now joined the ring ...

By 8 p.m., there have been 24 half-hour intervals, during which time the number of people joining the circle has been regularly doubling. The number of people who are in the know after 24 half-hours can be represented as the sum:

$$1 + 2 + 4 + 8 + \ldots + 2^{24}$$

What does this sum add to? How many people are now in on 'our little secret'? You probably suspect that it has run well into the thousands. In fact it is far worse than that. It turns out that, assuming every new disclosure was made to people who didn't know it already, there are now 33,554,431 people who have heard – roughly half of the UK population.

This phenomenal rate of increase is known as *exponential growth,* and defies most people's sense of how numbers work.

The final number reached by exponentials is also highly sensitive to how many people hear the news from a single source, what we will call the *reproduction number*. In our gossip example, each person was fairly discreet and told only two others, a reproduction number of two. If they had instead told three other people (which is still quite discreet), then, by six o'clock that evening, 5.2 billion people have heard – that's practically the *entire world population*. And you told only three people!

Spreading the gossip to new people doesn't necessarily mean everyone will get to hear about it, however. For it to grow, the reproduction number has to be more than one disclosure per person. If each person who hears the news spreads it to exactly one other person, then by 8 p.m. only 24 people have heard. The rate of spread in this case is steady and unspectacular.

If the news is sufficiently dull, or people are sufficiently good at protecting confidentiality, the reproduction number will be less than 1 per person. In this case, the story actually dies off. Suppose that out of a group of people who know the secret, three-quarters tell one person, and the rest keep their lips sealed, a reproduction number of three-quarters, or 75 per cent. Suppose, too, that 64 are in the room when the news is revealed. The spread goes like this:

8:00 a.m. 64 people know
8:30   75% of these 64 people pass it on (so 48 others hear about it)
9:00   Those 48 pass it on to 36 more
9:30   The 36 pass it on to 27 ... and so on.

The number in the gossip circle can be written as another series:

$$64 + (64 \times 0.75) + (64 \times 0.75^2) + (64 \times 0.75^3) + \ldots$$

This series can go on for ever. If you wait long enough, does this mean that the whole population will eventually get to hear? The answer is no. Only a limited number will ever get to hear it, and the spread of news will eventually stop altogether. In fact there is a formula for working out the sum of an infinite series like the one above, so long as the reproduction number, R, is less than 1.

If the number who hear the news at the start is A and the reproduction number of the gossip is R, then for the infinite series above:

$$\text{Total who hear the news} = A \big/ (1 - R)$$

In that example, A was 64 and R was 0.75. Plug the numbers into the simple formula to get: 64 / (1 − 0.75) = 64 / 0.25 = 256. That figure of 256 is known as the *asymptote*. It will never actually be reached, but when the number who have heard the news gets close to this figure, spread will cease.

Using the same formula you might like to confirm that if 200 people are given the news and the reproduction number is only one confidant per ten people (so R = 0.1), then only 200/0.9, or 222 people, will get to hear the gossip.

This says something interesting about news leaks. The formula shows that the number of people who eventually get to hear a news leak is far more dependent on the reproduction number than on the number who are exposed to it in the first place. Downing Street, take note.

The numbers behind infections

No doubt you can immediately see the similarity between the spread of gossip and the spread of an infection. The number of people who first hear the gossip is analogous to the number of people who are the initial carriers of an infection. The rate at which gossip is passed on to others is analogous to the infection rate of the disease. And, as we have seen, if gossip or a disease is going to become an epidemic, it is crucial that the reproduction number must be greater than 1. If that number can be kept below 1 – that is, if every carrier can be guaranteed on average to transfer the disease to less than one other person during the whole of their infection – then the disease will die out. This makes '1' probably the single most important number in the whole of epidemiology.

The reproduction number of diseases is dependent on a number of different things. The nature of the virus or bug itself is fundamental, of course. Some germs are so powerful and can infiltrate the body in so many ways, for example through touch or breathing, that they are highly infectious and hard to protect against. Others, such as HIV, are not particularly easy to transmit from person to person but may still have a high reproduction number because they survive for a long time and their carriers inadvertently behave in a way that gives the germ a helping hand in getting passed on, for example through the transfer of body fluid.

In order to work out the growth rate of the infection, all of these factors are taken into account by analysing the statistics of how rapidly the infections spread in human populations. In the box are some approximate figures quoted for the infectivity of four well-known diseases:

Reproduction numbers and infectious periods

	A typical infectious period	R number
HIV	4 years	3
Smallpox	25 days	4
Flu	5 days	4
Measles	14 days	17

In other words, at the start of an outbreak, a person with flu may be infectious for five days, during which time they will infect about four people. These figures are only rough averages, and depend on the specific virus, country and community in question. In developing countries the reproduction numbers are usually higher.

The key point is that all of the reproduction numbers are greater than 1, making all of the diseases a serious threat if they are left to their own devices. The rate for measles is particularly high, which is why it spreads like wildfire through classrooms of unimmunised children.

Why you need 'e' for natural growth

In the example of gossip-spreading earlier in the chapter, one assumption was that news spreads in regular clumps of half an hour. This is a gross simplification of what happens in reality. Infections don't wait till the clock passes a certain time before they have their next surge. They spread continuously.

There is a special number, known as 'e', that is behind continuous growth, and to understand this number it might help to think of it in money terms, as the box explains.

Where 'e' comes from

Imagine you have £1, and you put it into a bank account that offers 100 per cent interest per year. If the bank pays you the 100 per cent interest at the end of the year, you end the year with £2.

If, instead, it pays 50 per cent every six months, then you have £1.50 after six months, and 50 per cent more than that at the end of the year, or £2.25.

What about four lots of 25 per cent at three-month intervals? This works out at even more, a final amount of £2.44.

As the periods between interest payments get shorter, you get closer and closer to continuous growth of your investment, but the sum of money at the end of the year tends towards a maximum figure. That maximum is about £2.72. The actual number begins 2.71828 ... and is known as Euler's number, 'e'. It is the number at the root of all natural population growth and a fundamental player in many other areas of maths, too. Expressed as the formula $(1 + 1/n)^n$, the larger the value of 'n' becomes, the nearer you get to e.

Like the bank that adds interest continuously, infectious diseases constantly spread themselves – or, at least, that is a pretty close approximation.

The reproduction number, R*, is the number of new cases created by an infected person, and the number of people infected at the start of the outbreak is I. If infection only happened as a sudden event at the end of one infectious period, the number of newly infected people at the end of the period would be:

$$NEWLY\ INFECTED = I \times R$$

However, in the same way that the fictitious bank doesn't just add interest at the end, but adds interest on the interest, the new infections themselves start working on the population immediately. Not surprisingly, this leads to a formula that involves 'e'. The number of infected carriers after one infectious period (that's five days for flu or a month for smallpox) turns out to be:

$$\text{Number of carriers after one period} = I \cdot e^{(R-1)}$$

So if 10 people have flu at the start of the week and the reproduction number R is 4, this would predict that after one infection period (five days) there will be:

$$10 \cdot e^{(4-1)}$$
$$= 10 \cdot e^{3}$$
$$= 201\ people\ infected$$

---

* In epidemiology, the reproduction number at the start of an epidemic is known as R0. The zero indicates that this is the number of people that an individual will infect when the population has no immunity and is behaving normally.

After T infectious periods, the formula for the number of people infected is:

$$\text{Number of carriers after } T \text{ periods} = I \cdot e^{(R-1)T}$$

This is the fundamental formula of epidemics. If R is less than 1, then the expression $e^{(R-1)T}$ gets smaller as T gets bigger – in other words, the infection dies out. If R = 1, then the number of infected carriers remains constant. And, if R is more than 1, the infection becomes an epidemic.

This simple model of the growth of infection (and gossip) is quite accurate in the early stages. However, as more and more people become infected, there are fewer and fewer people left who are susceptible to infection. This must itself reduce the reproduction number. You can picture the analogy with gossip. After a while, it becomes harder and harder to find somebody who hasn't already heard the news, so the number of people to whom it spreads from each carrier will reduce. After a while, the pathogen begins to run out of people to infect, and if the rate of transmission drops low enough, the infection will die out before everyone has been exposed to it. In fact this is almost always what happens.

The Kermack McKendrick model

In 1927 two scientists, Kermack and McKendrick, developed a mathematical model that has become the reference point for all other major epidemic models since. They noted that the total number of infected carriers would grow if the number of new infections over a certain period was larger than the number of people becoming noninfectious in that period, the latter being achieved in one of two ways – by recovering or by dying.

Swine flu uncertainty

In early 2009, a new virus emerged in Mexico. It was officially known as H1N1, but its similarity to viruses that affected pigs earned it the nickname 'swine flu'.

As the virus began to spread internationally, there was a degree of hysteria in the press about a potential new plague, and epidemiologists began to make estimates of how many victims the disease might claim. There was a huge range in these estimates. Claims were made that the illness might kill anywhere between tens of thousands and tens of millions of people across the globe. The goalposts of these forecasts were set so wide that it's hardly surprising that the actual number of deaths fell in that range: it is estimated that perhaps 250,000 ended up dying from swine flu, a large number but thankfully only a tiny fraction of the worst-case forecasts. You might expect that after the event we should know exactly what the death toll was. One reason why the estimates of the real death count vary is the debate about whether to include those who died 'with' H1N1 but actually died 'of' something else, such as heart disease or a road accident.

But the method of counting doesn't account for the huge margins of error in the original forecasts. These were down to the sensitivity of exponential growth to the rate of infection. In the early stages of a new illness, the reproduction number R0 is little more than intelligent guesswork. As we've seen earlier in the chapter, a small margin of error in R0 can lead to a huge divergence in the end numbers, making it impossible to estimate the final death toll with any accuracy until the disease is well established.

KM, as we will call them, divided the population into three categories:

- Susceptible (i.e. haven't been exposed yet)
- Infected
- Removed (those who either recovered or died)

Their simplified model of the spread of infections took into account these three categories. The results proved remarkably good at replicating the patterns seen in real epidemics – a rapid growth in the number of people infected, followed by an equally rapid decline. The KM equations dealt with rates of change, and involved *differential equations*.

The mathematics of differential equations is certainly not trivial, and there is a severe risk that any further discussion on this topic will lead to a rapid glazing-over of eyes. So, instead, let's skip the formulae and jump straight to some of the things that KM found when they had solved their equations.

Most interesting is that the KM equations can predict the proportion of the population that will never be touched by the infection.

The greater the initial infectivity of the illness, the smaller the proportion of 'untouched' there will be. Remember that an initial reproduction number of more than 1.0 is critical for an epidemic to take hold. It turns out that, if the initial reproduction number is 1.5, more than 50 per cent of the population will never be exposed to the infection. However, as the reproduction number increases, the epidemic becomes more pervasive. By the time it reaches 3, only 5 per cent of the population remain unexposed.

Foot-and-mouth disease is an interesting case of this. Because it is so highly infectious, with an R0 of 50 or more, once one animal on a farm has become infected the *entire farm* is regarded by the modellers as being infected, and is treated as if it were one huge infected animal. Since the *infectivity* is considerably reduced at a mile separation, the spread of the disease is then viewed as being from farm to farm, with a more manageable infectivity rate of R at around 1.5.

With most infections, a critical starting rate is needed for the epidemic to take hold. Most often this means that people need to be packed together sufficiently closely, as often happens in the poor areas of towns, or the number of contacts between people has to be sufficiently high, through sexual promiscuity, for example. The consequence of this is that, if people can be kept far enough apart for long enough, most diseases will die out of their own accord with very few infections. This explains why the policy of lockdown was adopted across the world at the start of the coronavirus, and why the cruel policy of locking people into their own homes during the Great Plague was actually pretty effective.

Computer and other infections

As if biological infection weren't bad enough, mankind has imposed other viruses on itself of its own volition. Most notorious is the computer virus, which in many ways is a direct mimic of its biological cousin. Computer viruses are mini-programs written by programmers with a grudge or with too much time on their hands. The most powerful ones can wreak havoc on millions of computers by wiping out hard disks or clogging up email boxes.

Like living bugs, the computer versions can lie dormant before springing into action weeks or even years after they have entered your system.

There are, however, some important differences between computer and biological viruses. Biological viruses require physical contact of some kind, which means that the geographical location of susceptible people is important. You are much more likely to catch flu from your neighbour than from somebody in Bucharest. But, thanks to the Internet, geographical distance is no defence against computer viruses – they can travel from anywhere in the world in a split second. And, while biological viruses usually take hours or days to make their occupants ill, computer viruses can do equivalent damage almost instantaneously.

Finally, while mankind has a broad genetic diversity, making some people naturally immune to certain infections, computers increasingly lack this diversity (just think of how many people's computers contain identical Microsoft 'genes'). So, if one computer is infected, there is a risk that most might become so.

The consequence of all this is that computer infections can spread far faster than any viruses that have been witnessed before. They can infect tens of millions in hours instead of years, which makes them a major global threat to organisations of all kinds. A number of viruses have already caused this level of devastating damage. In May 2000, a virus wrapped up in an email headed LOVE LETTER FOR YOU was estimated to have reached 50 million users in just a week. Anybody opening the attachment suffered serious damage to their computer files, and many people did just that. One estimate put the damage at $2.6 billion.

The growing threat of computer viruses explains why the computer world has created roles that are direct parallels to those in the medical community. Scientists have produced mathematical models to predict the rate of infection and the risk of exposure from different computer viruses; computer 'doctors' are there to treat and with luck revive the infected.

And, most important of all, computer 'health' advisers are there to immunise or protect computers against infection; the golden rule is that prevention is always better than the cure.

The mathematics of infection applies even beyond the world of viruses. Very similar models are used to forecast the growth in the number of products in the market, for example. Marketing experts are forever looking at how they can increase the infectivity and penetration of their product, while minimising the infectivity of their rivals.

Religion, according to some, is also a virus. It tends to be passed on from parents to their children far more than between any other groups, and, while the infection can disappear in adulthood, it has a tendency to reappear in old age and other stages of life when immunity is low.

Even jokes and trivia spread with the same rules as viruses. Have you heard that St John's Wood is the only London Underground station that contains no letters in common with the word 'mackerel'? If you have, that's the virus at work. If you haven't, pass it on.

10

AM I BEING TAKEN
FOR A RIDE?

The formula behind the taximeter

In the film *Manhattan,* Woody Allen is in a taxi and quips to Diane Keaton, 'You look so beautiful I can hardly keep my eyes on the meter.' Taximeters have that sort of mesmerising effect on people. Yet, for all the hours people spend watching them tick upwards, few know the secret of how they work. Not even the drivers. When we asked a sample of London cabbies to explain the system, the response was typically, 'Good question, guv'nor, I've often wondered about that myself.'

To the passenger crawling through the traffic, one thing certainly seems clear. Whether you are hurtling through the back streets or stuck at the lights, the meter keeps ticking over. It seems that the taxi driver can't lose.

So what is the secret of the black box? Does the speed of your journey make any difference to the driver's pay packet? Can devious cabbies squeeze higher fares out of passengers by playing the system? These questions have occurred to all of us at one time or another, as another pound clocks up on the meter.

The basic formula

The basic principle behind a taxi fare is simple enough. If you make a long journey, you should expect to pay more than you do on a short journey, and the taximeter, a device invented by Wilhelm Bruhn in 1896, charges you a rate for distance travelled. But what about heavy traffic or unexpected delays due to an accident? As far as taxis are concerned, a 'long' journey means a long time, as well as a long distance. To cover the driver for his work time spent sitting in a traffic jam or crawling through the rush hour, the taximeter also has a rate for the *time* spent on the journey.

So a taxi fare is calculated using a formula that charges you for your distance and your time. Actually, that's a little misleading. It charges you for your distance or your time, but not both at the same time, as will become clear in a moment.

This distance-to-time approach is in fact the standard adopted internationally by taxis with meters, though it isn't the only model they could have used. In fact, if you think about it, the principle of charging you for travelling slowly is the opposite to the way you are charged for a train journey. On a train journey you don't pay any more for a journey that lasts twice as long.

On the contrary, these days, because of the system of refunds, the longer your train journey takes the less you pay!

There is a general rule in economics that the better the product is, the more it costs. But, if you think of a taxi ride as being 'get me there as quickly as possible', then the economic law here is the opposite: the worse the product, the more you pay.

Buried away in small print that few people ever bother to read is the formula for calculating this taxi fare. At the time of writing, the formula for the basic London fare for daytime travel looked like this:

£3 INITIAL CHARGE
+ 20p per 116.6 metres
or 25.1 seconds

Those suspiciously precise figures of 116.6 metres and 25.1 seconds are enough to put anyone off trying to work out the fare for themselves, though this isn't a deliberate ploy on the part of the taxi drivers to confuse passengers. The figure is actually set by the London authorities, and adjusted by an inflationary amount each year to try to keep driver earnings roughly fixed. If you go back far enough, there must have been a time when the distances and times were nice round numbers.

When the Secretary of State published the London Cab Order in 1934, the formula involved fractions rather than decimals. It looked like this:

3d (i.e. 3 old pennies) initial charge + 3d per ⅓ mile (if exceeding 5⅓ miles per hour) or 3d per 3½ minutes (if going at less than 5⅓ mph).

This figure of 5⅓ miles per hour was the estimated average speed of vehicles in London in 1934, perhaps kept low because of the relatively slow speed of horse-drawn vehicles, which were still reasonably common in the streets. What happened if your cab was travelling at exactly this average speed of 5⅓ m.p.h.? It turns out that, if you travel ⅓ mile at 5⅓ m.p.h., you will take exactly 3½ minutes, the same duration as the time unit. In other words, however you measure it, the cabby got threepence for each unit of the journey if he travelled at the average speed.

Before investigating the mysterious taxi formula any further, here is a little quiz. Suppose you take a black cab from your local station to your home every day.

Quiz

1. **Same distance, more time.** If your taxi journey today takes a couple more minutes than your journey yesterday, is your fare:
   (a) more than yesterday's?
   (b) less than yesterday's?
   (c) the same as yesterday's?

2. **Same time, longer distance.** If your journey today diverts around some back streets, adding a few hundred yards to the journey, but takes exactly the same time as yesterday's, is your fare:
   (a) more than yesterday's?
   (b) less than yesterday's?
   (c) the same as yesterday's?

Did you answer (a), (b) or (c) to either question? If you did, award yourself a point, because the answer to all of the options is 'Possibly'. Just because a journey takes longer in time or distance, it doesn't necessarily mean it will cost you more. Baffled? Read on …

How to calculate a city's average speed

If you ever want to know the average speed of traffic in a city, check the taxi fares. The fare will be quoted as X pence per Y distance and X pence per Z time. Divide Y by Z and you have a good estimate of the average speed. For example, in London Y and Z are 116.6 metres and 25.1 seconds; 116.6 divided by 25.1 = 4.5 metres per second, or just over 10 m.p.h. in New York Y/Z works out at 0.2 miles per 60 seconds, or 12 m.p.h. Why does this trick work? Because taxi rates are set by modelling what the expected income will be on an average day in average traffic.

Does the formula work?

The principle of the taxi fare is easy enough, but what exactly does it mean? Here is the formula used until recently by all New York cabs. It's the same principle as the London taxi, but the numbers are much simpler to handle:

$2.50 + 50c per 0.2 mile or 50c per 60 seconds (whichever is the higher)

One way to represent this formula is by plotting a graph:

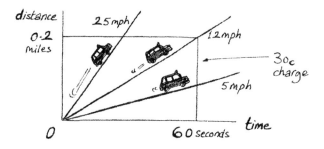

When you cross the rectangle on the graph, the taximeter clocks up 30 cents. At speeds above 12 m.p.h. – the 'critical speed' – it is the distance unit that clocks, but for slower speeds, it is the time unit. Once the unit has clocked the 30 cents, the counters reset back at zero. At the corner of the rectangle (which is reached by travelling at exactly 12 m.p.h.) the distance and time units clock simultaneously, but only one charge of 30 cents is counted.

This critical speed (12 m.p.h. in New York, but only 10.2 m.p.h. in London) is an important part of taxi fares. If your taxi travels faster than this speed, then for a journey over a set distance, the actual cost is fixed because you are charged only for the distance travelled. Below this speed the overall journey fare increases. Another graph illustrates this quite well:

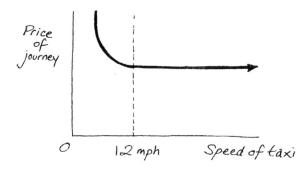

Towards zero speed, the cost of the journey soars upwards. In fact, if your taxi was parked permanently at a traffic light, the cost of your journey would head towards infinity – assuming you could be bothered to hang around that long, of course.

Notice how the curved part of the graph (for less than 12 m.p.h.) and the flat bit (over 12 m.p.h.) join each other. Smooth joins in graphs are a good way of avoiding fiddles – see page 132 for an example from everyday life where disjointed graphs encourage corrupt activity.

So far, so good. The formula looks reasonably innocent and fair. But beware, there are hidden traps. The best way to illustrate this is with an example that you could more or less imagine happening in real life.

Suppose you are with a group of friends in New York, and you are travelling from your apartment to a restaurant. You can't all squeeze into one cab, so you split into two. Both cabs leave at the same time, and arrive simultaneously at the hotel, having taken the same route. Imagine your annoyance when you discover that your friends' fare was lower than yours. How can this be?

To illustrate why, let's make the calculations simple. The cab journey is 1 mile, which takes six minutes. To start with the meter reads $2.50 which will cover the first segment of the journey.

Your friends' cab travelled at a steady speed for the whole journey, 10 m.p.h. Because this is lower than the critical speed of 12 m.p.h., the meter clocks for time rather than distance on this journey, adding 50c per minute. After six minutes you reach your destination and the fare is $2.50 + 6 x 50c = $5.50.

Now to your journey. Although the start and end time for your journey were identical to your friends', let's suppose that for the first minute, the driver of your cab put his foot down and drove at 36 m.p.h., covering 0.6 mile. Then you were trapped behind a slow vehicle and it took you five minutes to cover the remaining 0.4 of a mile.

This is how the taxi calculates your fare:

First 0.6 miles – 3 x 50c for each 0.2 miles = $1.50 as you were above the critical speed
Final 0.4 miles – 50c x 5 minutes = $2.50
Total cost of journey $2.50 + $1.50 + $2.50 = $6.50

In other words you paid \$1 more for your journey than your friends, despite the journey covering the same distance in the same time.

This example illustrates a quirk in taxi fares, namely that a journey on fast roads with frequent stops at traffic lights might well cost you more than a journey at a steady speed on slower roads. The longer the journey, the higher the discrepancy could be, and this discrepancy is possible in any taxi ride in any city, even if the standard taximeter has been perfectly calibrated. Because of this anomaly, it is possible to invent situations where the distance or the time of journey is bigger yet the fare is smaller (or vice versa). That is why the earlier quiz permitted every possible answer.

It would be hard for taxi drivers to exploit this deliberately, but, if they had a choice between smooth-flowing back streets or an expressway combined with slow exits, the latter might make them more money.

How does the cabby maximise his income?

Although the taximeter can be squeezed for a few more quid here and there, this isn't where the big margins are to be made as a cab driver. In the end, what he's most interested in is pounds per hour. The best way of achieving this is to have nonstop work, and to complete each job as quickly as possible. The time rate on the meter effectively sets the minimum wage for the driver. As long as he has a passenger on board, he knows he is earning at least 30c per 90 seconds (\$12 per hour) in New York, or 20p per 40.8 seconds (£17.60 per hour) in London. It may seem odd, by the way, to have a UK income that is double its US equivalent, but remember that this is only the minimum revenue, and among other things it probably reflects the higher fuel and training costs in London.

How taxes could learn from taxis

When pricing formulas are poorly designed, they can lead to undesirable customer behaviour. One example was the stamp duty imposed on property by the Chancellor Gordon Brown in 1999. The tax was as follows:

Price of house: £0–125K	*no tax*
£125K–250K	*pay 1% on top of the price as stamp duty*
£251K–500K	*pay 3% on top of the price as stamp duty*
£501K+	*pay 4% …*

This percentage was to apply to the whole amount. So somebody buying a house for £250,000 would only pay 1%, or £2,500, while somebody buying a house for £251,000 had to pay a 3% tax, or £7,530. The graph of stamp duty paid looked like this:

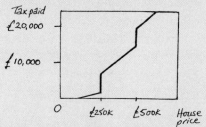

Understandably, somebody buying a house in the region of £251,000 would much prefer it to cost £250,000 than £251,000 because of the sudden £5,000 hike in duty. A similar disproportionate effect happened when the price climbs from £500,000 to £501,000. The sudden hike in tax caused a distortion in the market, with sellers finding all sorts of methods of arranging the price to be just below the boundary, some of which were more legal than others.

The moral from this is that graphs with sudden jumps can lead to corruption. That is why the smooth line of the graph of taxi fare against speed is a good one. It removes the incentive for a cabby to travel at a particular speed to receive a jump in income.

In 2014, Conservative Chancellor George Osborne changed the stamp duty rates, and at the same time removed the so-called cliff edges on the tax graph. It was one of his smarter moves.

But what is the ideal journey for a taxi driver – one that brings in the most pounds per minute? It turns out that there are two candidates – the very short and the very long journey. In both cases, they are favoured by a figure built into the fare formula.

As soon as a passenger enters a cab, he owes the driver a hire charge, £3 in London and $2.50 in New York. That represents a fantastic return for only, maybe, ten seconds of effort, so, in terms of income per second, passengers who are in the cab for a short time represent the best return. In practice, though, no passengers ride for less than a minute. So a second factor needs to be taken into account – the *tip*. Tipping can represent a significant proportion of a taxi driver's income. The most common sort of tip is to round up a fare to the nearest pound. This means that a cab fare of, say, £9.90, will often earn a total of £10, a miserly 1 per cent tip. The best fare to clock on the meter is probably something like £3.40. Chances are that the passenger will round this up to £4.00, a tip of nearly 20 per cent. Continuous £3.40 rides could bring the taxi driver something like £35 per hour. Nice work, if you can get it.

Ironically, high fares may also bring good tips. By the time somebody can afford a fare of, say, £42, they often have money to burn (their own or, better still, their employer's), so, according to one cabby in Aberdeen at least, a cheery 'Here's fifty pounds, – keep the change' is not out of the question. Another 20 per cent bonus. In addition, the distance rate on a taxi actually increases once the length of journey increases beyond a certain limit. This rule dates back to when taxis were pulled by horses and exhaustion was a consideration, but it also recognises that the further from the traffic hub a taxi is, the less chance it has of picking up a new ride very quickly. Nonetheless, there are anomalies. A long journey from a remote location such as Heathrow airport to central London not only

Taxicab geometry

According to Euclid, the shortest distance between two points is a straight line. But not according to a cabby. Because cities are laid out in grids, to travel from one point to another almost inevitably involves a zig-zag route. There is almost always more than one 'shortest distance' between two points. For example, in this grid, the shortest distance from A to B is five blocks, and you should be able to find ten different ways of doing it (one is illustrated).

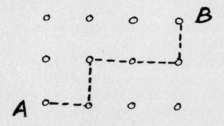

This way of working out distances has acquired a whole mini-branch of mathematics known as taxicab geometry. It has all sorts of curiosities. For example, below, the points marked X are all the same taxicab distance (two units) from the central point. And what do you call a shape where everything on the perimeter is equidistant from the centre? A circle, of course. So, in the world of taxicab geometry, you can square a circle – a trick that has eluded Euclidean geometers for centuries.

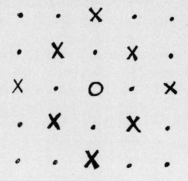

brings in a high rate of earnings, it also takes the driver straight to a hot pick-up zone. No wonder cab drivers love collecting newly arriving tourists.

In fact, whole mathematical models have been built to work out the optimal set of locations to send taxis to maximise the collection of passengers and income. There are also sophisticated models set up within the Department of Transport to look at traffic flows and establish what a fair fare should be.

As one cabby said, 'Blimey, you'd never have thought there was so much maths in it. And to think, I had that Rachel Riley in the back of my cab once.'

11

WILL I EVER MEET THE PERFECT PARTNER?

The chances and choices behind coupling up

It is well known that fewer people get married these days. Fifty years ago, getting married was expected of you, while remaining single carried a stigma, particularly for women. The term 'spinster', which has almost vanished from use, always seemed to carry far more negative connotations than its male equivalent 'bachelor'. The pressure was on to get married and have children as quickly as possible, and often people married their first love, and then stuck it out for better or for worse.

Those social pressures have reduced enormously, and people have begun to think much more of marriage as a lifestyle choice. As Bridget Jones put it: 'The whole bloody world's got a commitment problem. It's the three-minute culture. It's a global attention-span deficit.'

Why do some of us find it hard to commit to a permanent relationship? One of the many reasons is that we believe we have plenty of time to make a choice. Many of us set out to look for the 'perfect partner', whatever that might be, and put off engagement because we ask ourselves, 'What if the next person is even better?'

Making the best choice: The 37 per cent rule

Selecting a partner is an example of a *serial decision*. In other words, you aren't usually presented with all of your options at the same time, but instead they come along one after the other, and you have no idea who or what will come along next.

In fact there are similarities between finding a partner and other more mundane decisions, such as renting a flat, finding a parking space or accepting a job offer. In each case, the options are presented to you in turn, and, once you have rejected an option, you often don't have a chance to go back to it later. If you are driving down a one-way street, you can't turn back to take the parking spot you just ignored, and when the housing market is hot, then unless you say yes to the place you just saw, it is likely to be snapped up by somebody else. All are examples of serial decisions.

When is it right to stick with what you have? To examine this very real problem, it helps to have a slightly artificial case study, just to make the analysis simple. In this case, Jim is ideal. He is 39 years old, and is determined to be engaged by the age of 40. To remove some of the randomness from meeting possible partners, Jim has joined a dating agency, which guarantees to set him up with ten dates per year. We'll add the rather less plausible situation that every one of the dates he meets is so desperate to get hitched that all he has to do is pop the question. Jim therefore knows that one of the ten dates that he will meet will become his wife. But which one?

It seems a little callous to rank possible partners in order, but unfortunately this is necessary if the analysis is to be taken any further. (And it has to be said that men and women are not averse to rating potential partners in this way.) One of the ten dates will be the best choice of partner and one is the worst, but the order in which Jim gets to meet them will be completely random.

Jim meets his first date, and she seems OK. But is she the best one? She might be, but the chance of her being the best of all the dates he is going to meet is only 1 in 10. So it seems a reasonable decision not to commit to her, but to use her as the benchmark with which to compare later dates (we're avoiding any judgment on Jim's moral character here).

If he's a real ditherer, Jim could continue being noncommittal by saying no until the final of the ten comes along, at which point he has no choice – he has to go with her if he is to become engaged as he promised. If he opts for this dithering, noncommittal strategy, he still has only a 1-in-10 chance of picking the best. There must be a better strategy.

Indeed there is. One way to improve his chances is to say no to the first date, whom we will call Kate, but say yes to *the first date he meets who scores higher than Kate*. Using this strategy, nine times out of ten he will end up with a better partner than Kate. However, if Kate happened to be the best date, this strategy fails.

Choosing the first date who beats Kate actually increases his chances of ending up with the best possible partner to 20 per

cent, or 1 in 5. The calculation of this is quite tricky, since it is necessary to add together the chance of the best date arriving second, or third (with the second scoring less than Kate), or fourth (second and third both scoring less than Kate) and so on up to tenth.

What if Jim uses more than just Kate to be his benchmark? The longer he holds out, the better a cross-section he has of all the possible dates, but the more chance there is that the best partner will already have been rejected.

In fact, there is mathematically a best solution for Jim in the above circumstances. This is to meet three dates, and then to propose to the first one who improves on all of them. This way Jim's chances of ending up with the best available partner increase to about 1 in 3. You can simulate Jim's experiences by playing the blind-date game described in the box.

Despite its gross over-simplification of real life, this exercise makes a good case for what many people do anyway, which is to experience a few partners before making a firm commitment. As the number of potential partners increases, the mathematical answer tends towards an extraordinarily precise ratio – if there are N partners available, you should commit after meeting N divided by 'e' partners, where 'e' is roughly 2.718, and is the number at the heart of exponential growth (see Chapter 9). When N is a larger number, this means you should settle down after meeting about 37 per cent of your potential partners.

One of many flaws in this cunning plan, of course, is that you have no idea beforehand how many potential partners you will meet. Still, if you estimate that you might expect to meet forty potential mates in a lifetime, then after meeting fourteen of them it is time to look to settle down.

How commitment-phobic are you? Play the blind-date game.

Pick out ten cards from a pack that are numbered ace to ten, with ace scoring low. These cards represent your blind dates, and the aim of the game is to end up with the highest-scoring card. Shuffle the cards, and deal them out face down:

Starting at the left, you can choose how many of the cards you want to date, but not get serious with. This is you, 'playing the field' without making a commitment. The lowest playing-the-field (or PF) number is zero – in other words, you'll take the first card that comes along. The highest PF number is nine, which means your commitment is at a minimum but you are forced to commit to the last card you turn up, the tenth. Once you have chosen your PF number, turn over that many cards, and note the highest score. This is your benchmark. Now start turning over your potential partners. The first card that is higher than the benchmark is your partner. If none scores higher, the tenth card is your partner regardless of its value.

On average, PF numbers of 0 or 9 give you the worst outcomes. The best results come from choosing a PF number of 3. This will give the best available partner about one time in three.

The perfect match?

The strategy just described gives you a good chance of meeting the best possible partner of *the ones you are destined to meet*. This isn't quite the same as meeting the perfect partner. For example, if your main interest in life is travelling to remote countries, you're going to be disappointed if none of the people you meet has a desire to travel any further than Clacton-on-Sea.

To find somebody truly compatible requires a more focused search. This is where dating agencies can come into their own.

By asking lonely hearts to fill in a questionnaire, they can then use the data to match up possible couples based on their interests and preferences. Statisticians like to use *distance measures* to find the degree of match between two sets of data. The principle is that the smaller the overall gap between the two sets of data, the better matched the two items will be.

As an example, consider Annie, who is seeking a partner, and fills in a short questionnaire. For questions with yes/no answers, she puts a 1 for yes and a 0 for no. Here are her answers, together with those of two possible partners, Ken and Josh:

	Annie	Ken	Josh	A:K	A:J
Prefer city to country?	1	0	1	1	0
Prefer a night in to a night out?	0	1	0	1	0
Like cats?	1	0	1	1	0
TOTAL				3	0

The final two columns show the distance between Annie and Ken and between Annie and Josh for each factor. Annie/Ken have a total distance of 3, while Annie/Josh have a distance of 0. This makes Annie and Josh a perfect match according to this simple test.

Not surprisingly, however, this simplistic approach is flawed. What happens if the questionnaire includes questions with numerical answers that are not in the range 0 to 1, such as age?

	Annie	Ken	Josh
AGE	29	31	35

Added to the previous result, Annie's distance from Ken is now 2 years + 3 points from the other questions, making 5, while her distance from Josh is 6 years + 0 points, making 6. Since Ken has the lower distance, he is now deemed the more compatible by this scoring system.

There is something wrong with this. Annie is completely compatible with Josh on preferences, but this match is wiped out by their age difference, which is on a completely different scale. A better distance measure would make sure that each category was scaled to give it a similar degree of variability. The fact that Josh is six years older than Annie may be worth, perhaps, two points of difference, rather than the six shown above. Even yes/no questions may not all be on the same scale. If Annie's main passion in life is cats, then it might be more appropriate to score your liking for cats on a scale of 0 to 5, so that it outweighs other less critical factors.

There is another potential problem with questionnaires, too. Here are some more of their answers:

Do You LIKE:	Annie	Ken	Josh
Watching TV?	O	1	1
clubbing?	1	O	1
hip hop?	1	O	1
garage music?	O	O	1
pop?	1	O	1

Here Annie and Ken score a difference of 4 (they only match on garage music), while Annie and Josh score 2, so once again Annie and Josh are the more compatible. However, four of the five questions are closely related to each other. An interest in clubbing will tend to mean an interest in the various types of music heard in clubs, too. Since Annie and Josh both like clubbing, this is bound to have distorted the scoring in Josh's favour.

It would be so much better if there was a statistic that could reduce the distorting effect of questions that were closely related to each other, such as the clubbing and music ones above. Sure enough, a statistic exists that does this and adjusts the scaling to remove anomalies like the age differences earlier. It is called the *Mahalanobis distance,* based on the principles that have just been discussed but calculated using an intimidating set of vectors and matrices that will do nothing but confuse matters if reproduced here.

Mahalanobis is widely used in the field of statistical matching. For example, companies wanting to know when to advertise their products on TV will benefit from information that shows them which type of programme is watched by the sort of people who buy their product. A bit of *data fusion,* courtesy of Mahalanobis, will give them a good indication.

This approach would be great for use in the databases of dating agencies too – except for two things. The first is a tendency to distort the truth when describing oneself, which means that the distance calculation will only be finding the compatibility with the person you *say* you are. That might be very different from the real you.

The second is even more critical. There is an old saying that opposites attract. If there is any truth in this then the whole argument about matching by finding the shortest distance goes out of the window.

Maybe this is why most dating agencies don't bother to use the data-matching techniques that are used by other database industries. Instead, they tend to leave it to chance and let nature take its course.

Competing for partners

Finding a suitable partner is a tough enough problem when there is no competition. The complexity increases enormously when a whole society of people are jostling to find their ideal partner. What's the chance that people will find a partner, and what chance is there that, having found one, they will be happy?

Sociologists, counsellors and psychologists have devoted much thought to this problem, and mathematicians have also had their say.

In fact the stable-marriages problem was analysed with notorious success by two mathematicians called Gale and Shapley back in 1962. They came up with a series of instructions – known as the Gale-Shapley algorithm – for setting up couples that guarantees a high level of satisfaction, at least for one half of each couple. All that is required for the model to work is the same number of males and females, with a fundamental assumption that all are seeking a partner of the opposite sex. We will stick with tradition and assume that it is the men who propose to the women, though everything that follows could of course apply in reverse. The routine goes like this:

- Each man in turn approaches the woman who is top of his list and proposes to her.

- If she has nobody with her at the time, she says 'Maybe', and the man stays with her.
- If she already has a man with her, she then chooses between the two, saying 'No' to the one she likes less. The rejected man then approaches the next woman on his list.

This continues until every man has found a woman who does not reject him. It may sound Victorian, but this process does at least ensure that every man ends up with the best woman who is willing to have him. The men can therefore be said to have achieved the best outcome they could have hoped for. Unfortunately things are not nearly so rosy for the women, who end up in what is usually a far from ideal arrangement from their point of view.

Take the following example, with three men – Andy, Brett and Charles – being paired up, most conveniently, with Xenia, Yvonne and Zoe. The order of preference is shown for each of them:

ORDER OF PREFERENCE			
ANDY	X	Y	Z
BRETT	Z	Y	X
CHARLES	X	Z	Y

ORDER OF PREFERENCE			
XENIA	B	C	A
YVONNE	B	C	A
ZOE	A	C	B

So, for example, Andy's order of preference is Xenia, then Yvonne and lastly Zoe.

In the first round, Andy and Charles both choose Xenia, while Brett joins Zoe. This leaves Xenia facing a choice between Andy and Charles, and she opts for Charles because he is higher in her list. Andy, now rejected, therefore opts for his second choice, Yvonne. Each person has now become part of a couple.

The couples have ended up as:

Andy-Yvonne
Brett-Zoe
Charles-Xenia

Brett and Charles both got their first choices, while Andy ended up with his second choice. No wonder they are smiling. But their partners don't look nearly so happy. None of them got their first choice, and in fact Yvonne and Zoe both ended with their bottom choice.

This approach produces a different outcome if it is the women who are doing the proposing. In this case, going through exactly the same process but with the women in the box seat, the couples end up as:

Xenia-Charles
Yvonne-Brett
Zoe-Andy

Now it is the *women* who are happier, with Yvonne and Zoe paired with their first choice male and Xenia landing her second choice.

In fact, the Gale-Shapley algorithm will always favour the proposers. Whatever else you might think of it, one benefit of the approach is that the marriages it produces are 'stable'. That is, however much one partner may want to split up, no other partner will prefer the would-be divorcé to the partner he or she already has.

As with so many mathematical theories, once the algorithm had been applied to marriages, people began to see applications to all sorts of other areas where partners were being brought together. For example, what about qualified doctors applying to work at hospitals, or prospective tenants looking for a house to lodge in? By following Gale-Shapley's rules, whatever the final arrangement, one party will be as happy as it can be, although there may be some squabbling over which side gets to be the proposer, since overall they will be the winners.

There have been attempts to improve on this algorithm, to come up with solutions that are closer to the ideal for both sides. Unfortunately, real life throws up complications that limit the effectiveness of even the cleverest systems. One problem is that some people refuse to accept second best. Once their 'ideal' partner has turned them down, they would rather stay single than commit to anyone else. Even worse, however, preferences change over time. The ideal partner of today may be very different from the ideal partner in ten years' time. Stable partnerships therefore become unstable.

It would seem that the ideal system for identifying the perfect partner needs to consider a number of different factors. It needs to be able to detect genuine signals from distorted ones. It needs to be able to forecast who else is going to turn up. And, most of all, it needs to have the power to forecast how people will change. Needless to say, the perfect system for finding the perfect partner hasn't been found yet.

12

IS IT A FAKE?

Number tests that can detect the fraudsters

Criminals can inadvertently leave all sorts of clues about their crimes, such as fingerprints, fibres of clothing or weapons. But there is another less tangible type of clue that can prove to be just as incriminating. In a whole range of activities, from business to the laboratory, a fraudster has been discovered because of the *numbers* he has left behind. Not phone numbers or bank accounts, but ordinary-looking, everyday statistics.

One of the most curious pieces of evidence that indicates potential fraud hinges on the number 1. To understand the principle behind it, take a look at the front page of today's newspaper. Almost certainly lots of numbers will appear on it, in all sorts of contexts. For example: '50,000 more troops …', 'cut by 2.5 per cent', '… last happened in 2002 …', 'he gave 18 precise instructions', '… father aged 65 …', or 'continued on Page 3'.

These numbers are all completely unrelated to each other, but do they display a pattern? What proportion of the numbers in a newspaper would you guess begin with the digit 1? What proportion begin with 5 or 8?

You may never have given this any thought, but it would be natural to suppose that the first digits of numbers in a newspaper are pretty evenly spread. In other words, you might expect that a number drawn randomly from a newspaper is as likely to begin with a 1 as with a 9.

Surprisingly, this is not the case. In fact a number drawn randomly from the front page is much more likely to begin with a 1 than any other digit. Almost half the numbers will begin with a 1 or a 2, and the larger the digit the less likely it will be to appear at the start of a number. Numbers beginning with 9 are quite rare. If you collect enough results, you should find something close to this pattern emerging:

Leading digit	Chance of appearing
1	30%
2	18%
3	12%
4	10%
5	8%
6	7%
7	6%
8	5%
9	4%

How can these numbers be predicted so accurately, when the newspaper numbers are drawn from an unpredictable selection of stories? This odd distribution of numbers is determined by what is popularly known as *Benford's Law*. In 1939, Frank Benford, an engineer at GEC, made a curious observation. When he was looking at statistics for the populations of cities, far more of the numbers began with '1' than with any other digit. He investigated this further, and discovered that it also applied to stock prices, river lengths, sports statistics – in fact almost any collection of everyday numbers.

Benford's Law turned out to work everywhere, as long as the sample of numbers was sufficiently large, and the numbers concerned were not constrained by some sort of rule or by narrow limits. Telephone numbers do not obey Benford's Law, for example, because they are constrained by having to be seven or eight digits long. The height of adult males would also not fit the pattern because males are almost always between 150 and 180cm tall. As long as these conditions are borne in mind, however, the Law is remarkably powerful.

Benford's Law and fraud detection

Early in the 1990s, Benford's Law made its famous entry into the world of fraud detection. Mark Nigrini, a lecturer in accountancy, asked his students to look at the accounts of a business they knew, in order to demonstrate to themselves the predictable distribution of first digits. One student decided to look at the books of his brother-in-law, who ran a hardware shop. To his surprise, the numbers didn't resemble the Benford distribution at all. In fact 93 per cent of them began with the digit 1, instead of the predicted 30 per cent. The remainder began with 8 or 9.

The discrepancy was so huge, that it suggested that something must be wrong with the figures. In fact, rather embarrassingly

The formula for Benford's Law – and why it works

Proving Benford's Law is hard, but here is one way of seeing why it might be true.

Imagine you are setting up a raffle, in which you will randomly draw a number out of a hat. If you sell only four raffle tickets, numbered 1, 2, 3, 4, and then put them into a hat, what is the chance that the winning number will begin with 1? It is 1 in 4, of course, or 25 per cent.

If you now start to sell more raffle tickets with higher numbers 5, 6, 7, and so on, your chance of drawing the 1 goes down, until it drops to 1 in 9, or 11 per cent, when nine tickets have been sold. When you add ticket number 10, however, two of the ten tickets now start with a 1 (namely 1 and 10), so the odds of having a leading 1 leap up to 2 in 10, or 20 per cent. This chance continues to climb as you sell tickets 11, 12, 13 ... up to 19 (when your odds are actually $^{11}/_{19}$, or 58 per cent). As you add the 20s, 30s and above, your chances of getting a leading 1 fall again, so that when the hat contains the numbers 1 to 99, only $^{11}/_{99}$, or about 11 per cent, have a leading 1. But what if you put in more than 100 numbers? Your chances increase once more. By the time you get to 199 raffle tickets, the chance that the first digit of the winning ticket will be 1 is $^{111}/_{199}$, which is over 50 per cent again.

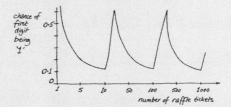

You can plot your chance of winning this game on a graph. On the vertical axis is the chance that the number you draw will begin with a 1, and along the bottom is the number of raffle tickets sold.

Interestingly, the chance zig-zags between about 58 per cent and 11 per cent as the number of tickets sold increases. You don't know how many will be sold in the end, but you can see that the 'average' chance is going to be somewhere around the middle of those two, which is what Benford's Law predicts. The exact chance that a number will begin with digit N as predicted by Benford's Law is: log (N+1) – log (N), where log is the logarithm of the number to base 10 (the log button on most calculators). For N = 1, this predicts log (2) – log (1), or 0.301, which is 30.1 per cent.

for all concerned, the student had inadvertently discovered that his relation had been fraudulently cooking the books.

From these small beginnings, Benford's Law has grown to become one of the formal tools that many accountants use to try to detect fraud. The joy of the method is that it is so simple to use. Just occasionally it comes up trumps, as in one case in Arizona, where the accused was found to have made too many cheque payments that began with the digits 8 and 9. The numbers themselves looked innocuous enough, but they were completely at odds with the downward curve predicted by Benford. This pattern was actually typical of people who are inventing fraudulent figures. Often they will invent sums of money that are just below a significant threshold, such as £100, where further authorisation may be required. In doing so they are distorting the natural pattern of numbers, and leaving clues for investigators that something odd is going on.

What if the statistics are *too good?*

Fraudsters don't just operate in business. Figures can be cooked in all sorts of places, particularly in science. Scientists are always under pressure to come up with results that produce the answers they, or their sponsors, are looking for. Often these will be newsworthy discoveries, especially findings such as a new wonder drug. The temptation to give the statistics a helping hand must sometimes be overwhelming.

This is not just a modern phenomenon. In the 1950s, the psychologist Cyril Burt was keen to find out whether intelligence was mainly determined by your genes or your upbringing. To use the modern expression, he was testing nature against nurture. To do this, he tracked down identical twins who had been separated in their infancy, so that he could compare their performances in intelligence tests. Because they were identical, they had the same genes, but they had experienced very different upbringings. As a comparison, he found a group of non-identical twins who had been brought up together. These had different genes but almost identical upbringings.

The statistical test that Burt used to examine these groups is known as a *correlation coefficient*. This is a statistic that measures how closely two results vary with each other. For example, the correlation between the temperature outdoors and the consumption of ice cream is probably quite high – in hot weather, lots of people buy ice cream, whereas on cold days there is very little demand. On the other hand, there is probably no correlation between the sales of ice cream on a particular day and, say, the number of children born that day in Liverpool. The two statistics are completely independent of each other.

In the case of the twins, if genes were the main influence on intelligence, Burt would have expected the separated identical twins to have a high correlation in their IQ tests. In other words, if nature determines your intelligence, it doesn't matter what kind of home or school you go to, your brightness will shine

153

through. However, if upbringing was more important, then the nonidentical twins brought up in the same home should have had the greater correlation.

Burt's results showed a much higher correlation for the identical twins living apart. The actual result was a coefficient of 0.771 out of a maximum possible of 1.0. This was very high, and seemed to provide strong evidence that it was genes that mattered most.

What raised suspicions, however, was that in a later experiment Burt confirmed his earlier result by finding, once again, a correlation for identical twins of 0.771. This could, of course, have been a coincidence, but the investigators didn't think so. Scientific results always have a certain amount of random fluctuation, and the chance of producing the same result to three decimal places was almost certainly less than 1 per cent. Taking this and other factors into account, five years after his death Cyril Burt was found guilty of fraud by the British Psychological Society. To this day it is still not certain if this conclusion was fair, but without doubt it demonstrated that if you fake your results, it pays not to be *too* consistent.

Who leaked to the press?

A different form of cheating is the anonymous leaking of confidential documents to the press. This must be hugely annoying for the authors, at least when they aren't the ones responsible for the leak.

To help trace the source of leaks, one software company several years ago devised an ingenious method of labelling documents. When the originals were printed out and distributed, each document appeared to be identical. The person leaking the document could therefore feel confident that it could be passed on without any clue as to who the culprit was. However, the word processor's software was adapted so that on the bottom line of one page, the spacing between the words was different in every document. For example, the first copy could say:

This will almost   certainly lead to an increase in unemployment.

The second would say:

This will almost certainly lead   to an increase in unemployment.

The two might look identical, but, if you examine them more closely, you will see that the first has a wider space between the words *almost* and *certainly,* while in the second the gap is between *certainly* and *lead*. Spaces are being used as a unique, secret code to identify the recipient of each document. In fact this is a form of binary code. In the above sentence there are ten words and nine spaces between the words. If a normal space stands for 0 and a double space for 1, then the first sentence has the code:

001000000

And the second is coded:

000100000

Using a string of nine 0s or 1s, there are altogether 512 different combinations that can be made using just that one innocuous sentence -plenty enough to ensure a unique code for everyone on the distribution list of most documents.

If the leaked document can be recovered, the source of the leak can be tracked down immediately. Who knows how many documents coded in this way you have received over the years?

Did Shakespeare fake it?

Detective work also extends to the world of literature. Did Shakespeare really write all of the plays that today carry his name? Scholars continue to debate this, and the more numerate of them have used statistical methods to investigate the authorship of certain works.

How do you statistically analyse a work that is supposedly by Shakespeare?

The simplest approach is to count how often Shakespeare used certain words in those works that are known to have come from his pen. Certain words appear frequently, such as the words 'world', 'naught' and 'gentle'. Other words never crop up at all, such as the word 'bible' (this odd fact often crops up in pub quizzes). If the work under investigation included the word 'bible', suspicions would immediately be raised that this was not a Shakespeare piece, and in fact the relative frequencies of the different words can be directly compared to see if they follow the familiar pattern.

Investigations of authorship do get much more sophisticated than this, however. In 1985, a poem entitled 'Shall I die?' was discovered in the Bodleian Library of Oxford University. On the manuscript were the initials W.S. Could this be a forgotten Shakespeare work?

The detection of exam cheats

A different but equally important type of fraud has been known to take place in the classroom or exam hall. Every year, a school pupil or student is accused of cheating, often by copying from a neighbour in some way. Like all other fraudulent acts, it is usually a statistic that stands out from the norm that first arouses suspicion. In this case, the abnormal statistic is usually a student who performs much better than his predicted grade.

At this point, all sorts of investigations can be made, including comparisons between the suspect's answers and those of the neighbouring students. Usually maths is not needed to detect the guilty. The same wrong answer using identical wording on two neighbouring papers is incriminating enough. For multiple-choice tests, however, it would be relatively easy to devise a method comparing the selections made by two neighbouring candidates. Incorrect answers are likely to be the key. The chance of both candidates getting consistently the same wrong answers is extremely low.

The investigations began. One early analysis was based on the patterns of words that Shakespeare used as his career progressed. In each new work, it turned out, Shakespeare had always included a certain number of words that had never appeared in any of his earlier works. (Fortunately, computers are able to do all of the word counting to prove this. Imagine how tedious this sort of analysis was before the electronic age.)

It was therefore possible to predict how many new words might be expected in a new work. If there were too many, it would be pretty clear that the author couldn't be Shakespeare. No new words at all, and it would look suspiciously as though somebody had tried too hard to copy Shakespeare's style.

The mathematical prediction was that the poem 'Shall I die?' should contain about seven new words. In fact it contained nine, which was pretty close. This was used as evidence to confirm Shakespeare's authorship.

But the sceptics were not convinced, not least because the poem didn't read like a Shakespeare. Many other analyses were carried out on the words. One devoted professor looked not at individual words but at the connections between words. As an example to show how this could work, two authors might both use the words 'heaven' and 'earth' equally often, but one might always use the two together, while the other always uses them individually. Each pattern will be a distinctive feature of the author. Indeed it might be said that the words resemble a DNA sample or a fingerprint, though this analogy has to be treated with great care because, while a person's DNA never changes, the patterns in their text might vary considerably from document to document.

The word-connection test appeared to rule out Shakespeare, though unfortunately it ruled out all of the other leading contenders too, such as Marlowe and Bacon. This test didn't convince everyone, either, and in fact the debate continues about who really wrote the sonnet. Current opinion seems to favour Shakespeare as author, but it does depend on which test you trust.

There is in fact a whole range of different statistical tests that can be carried out on documents. Others include average sentence length and average word length. Even parts of words can be compared, for example by breaking up a document into five-letter chunks and performing a huge number-crunching analysis on the frequency and distribution of patterns.

One of the highest profile cases that used this approach hit the news in 2013 when The Sunday Times revealed that the author Robert Galbraith was in fact none other than J.K. Rowling writing under a pen-name. In researching this exposé, they had sought the help of two linguistic scholars, Peter Millican and Patrick Juola. The experts' approach was to compare Galbraith's novel with the work of four female authors, including J.K. Rowling, Ruth Rendell and P.D. James. One comparison was of the distribution of word length. Of the eleven sections of text they examined, more than half were closer to Rowling's style than to the other authors. In a test of 'bigrams' (pairs of adjacent words), nine of the eleven resembled Rowling's style more closely than the others. None of this was proof, but it certainly suggested Rowling was a likely candidate – and not long after the article was published, Rowling 'fessed up to her pseudonym.

Does this help in crime detection? It can at least provide supporting evidence. On a couple of occasions, most notoriously in the case of the Unabomber in the USA, letters sent by the suspect have been compared against other documents he has produced to see whether the patterns match. However, to be truly confident of identifying the author of a work purely by the words used, as opposed to handwriting style or spelling mistakes, it may be necessary to have thousands rather than hundreds of words in order to get a conviction. In other words, it takes more than a sentence for a judge to give a sentence ...

How chi-squared secured a date

One statistical test used by literary analysts, as well as by scientists and many others, is known as the chi-squared test. In this test the observed frequencies in the sample, such as the number of occurrences of words such as 'bible' and 'discontent', are compared with the expected number. The conclusion of this test is expressed as a percentage chance: for example, 'Fewer than 5 per cent of Shakespeare's documents might be expected to have produced the patterns we see here.' In one of its more unusual applications, back in the 1980s a student friend of ours used a chi-squared test to demonstrate to manufacturer Nestlé that the letters inside the caps of packets of Smarties chocolates were not randomly distributed. He had been trying in vain to collect the letters of his Valentine's name. He won the argument, received some free packets – and got another date out of it, too.

How many frauds are there out there?

No detection system is perfect, and some frauds are always bound to slip through the net. Even though they may not be found, it is actually possible to make an estimate of the number of fraudsters who get away with it, based on the number that are found.

The technique for calculating the number of fraudsters would be the same as one that is used by proofreaders checking a document. Everyone knows that typographical errors (called typos by those in the trade) are sometimes difficult to spot, so a printer might ask two proofreaders to read through independently to look for errors.

Suppose the first reader finds El errors and the second finds a different number, E2. They now compare their results, and discover that some of the errors, a number S, were the same ones. How many errors might they expect there to be in total?

There is a way of making a good estimate, known as the *Lincoln Index*. This says that the total number of errors in the manuscript will be roughly:

$$\text{Expected errors} = \frac{E_1 \times E_2}{S}$$

For example, suppose the first reader found fifteen errors and the second twelve, and that ten of the errors were found by both. The Lincoln Index predicts 15 x 12/10 = 18 errors in total. Of these only seventeen have been found so far – ten found by both readers plus five more that only the first reader found and two more that the second found.

Exactly the same technique could be used by, say, tax inspectors to estimate how many fraudulent tax forms were getting through. Two inspectors would independently check through a pile of forms, and identify the suspect ones. If the first clerk found 20 and the second 24, with 12 in common, then the inspectors have identified 32 different suspect claims between them. The Lincoln Index suggests that there are 20 x 24/12 = 40 suspect forms in total. This means that about eight have slipped through.

As far as we know, this has never been used as a fraud-detection technique, but it would be an interesting one to test out. In fact, added to the other devices described earlier, it's enough to make amateur sleuths of us all.

13

WILL THE UNDERDOG WIN?

The maths behind memorable sporting moments

Everybody has a favourite sporting moment. Maybe it was Jessica Ennis winning heptathlon gold on Super Saturday at the 2012 Olympics, Andy Murray in 2013 becoming the first Brit to win the Wimbledon men's singles since 1936, or England winning the cricket world cup in 2019.

How the TV executives long for such thrilling moments, since this is what causes sudden surges in the number of viewers. The sports authorities love the big moments, too, and over the years they have not been averse to the odd rule change here and there to improve the chance of more thrills in a shorter time interval.

What, then, are the key factors that make a thrilling sporting moment? If you trawl through all of the great moments, certain themes seem to crop up time and again.

Victory for the underdog

Crowds – especially English crowds – love to see the little guy win. The nation was gripped when, in 2016, Leicester City won the Premier League having being given odds of 5000:1 against at the start of the season. Every year, too, an unknown Brit upsets a stronger player at Wimbledon before disappearing into oblivion. And in 1997, a little-favoured European Ryder Cup golf team managed to overcome the mighty Americans.

Although there don't seem to be any hard statistics on this subject, it does seem that some sports throw up more victories by 'underdogs' than others. There are many examples of the proclaimed underdogs winning in football, and examples are also quite common in cricket, tennis and golf. On the other hand, in rugby, athletics, rowing and many other sports it is unusual for the outsider to win.

Before we see why this might be the case, we need some definition of what is meant by an underdog. Since underdogs aren't given much chance of winning, you might define them as competitors to whom the bookmakers give only a slim chance of winning – less than 10 per cent, perhaps. However, this simple definition doesn't work if there are more than two competitors. Even the favourite in the Grand National only has about a 10 per cent chance of winning.

In any case, it seems wrong to use probability as a way of defining an underdog. Suppose we defined an underdog as a competitor with only a 1 per cent chance of winning. We would expect that the number of underdog victories would be the same in all sports (i.e., they would win 1 per cent of the time). The fact is, however, that underdog victories happen far more often in some sports than in others.

So instead, we won't define underdogs by their chance of winning, but by their relative 'weakness' compared with their opponents. Some sports are merciless on the weak, who don't stand a chance. Others, however, thanks to the scoring format or the opportunity for flukes, are much more favourable. These are the sports where underdogs are more likely to come out on top.

Tennis is a good example of a sport where the most talented player doesn't get all the credit he or she deserves. If you really wanted the technically better player to win a tennis match, then you would have a scoring system such as 'first to reach 100 points wins'. Bjorn Borg and Serena Williams at their peak would have been unbeatable. But this would make for very dull sport (see the later section on 'lead-swapping').

Instead of this, the hundred or more points in a major tournament match are broken down into 'big points', more commonly known as sets. Regardless of whether a player wins a set by six games to love, or by seven games to six, they still only get one set. Because of this, there are many examples in tennis where the player who won fewer points won the match. In fact, it is hypothetically possible for a player to score almost twice as many points as the opponent, and still lose. Suppose Andy Murray lost a match 6-0, 6-0, 6-7, 6-7, 4-6. If you are familiar with tennis scoring, you can figure out the extreme numbers of points he could have won and lost in each game and the match overall. It's possible for the loser to have 'won' this match by a whopping 158 points to 86. Is this the biggest margin of 'winning' defeat possible in any sport?*

Underdogs are also favoured in sporting contests where there are very few scoring opportunities. One of the lowest scoring of all sports is football. A typical match has only two or three goals. There are, however, far more scoring chances than goals. Although it is simplifying matters a little, there is some truth in the argument that, while the number of scoring chances is a direct reflection of the team's skill, the number of chances that are converted into goals is more a matter of luck. Everton might therefore 'beat' Peterborough by fifteen chances to four, but, if the probability of a chance going in the net is only about 1 in

---

* Some wise-cracker once pointed out that in tennis the winner actually has to win only one point – the last one.

5, this means an expected result more like 3-1. The winning margin is just two goals, and, when margins are this small, random variation says that there is a decent chance of this becoming a 2-2 draw or even a 2-3 defeat.

There is one other major factor that can benefit an underdog, and that is the freakish accident or fluke. In 1967, a horse called Foinavon won the Grand National, despite being the rank outsider at 100-1 against. Had the horses all completed the course, Foinavon wouldn't have had a chance. On this occasion, however, about twenty horses fell or refused at one fence. Foinavon was so far behind the rest that he was able to avoid the melée and had a free run for victory. In this case being hopeless turned out to be a huge advantage to the underdog.

Motor racing is also prone to accidents or car problems that randomly hit the strong as much as the weak. Almost always the reason why a non-favourite wins a grand prix is because something other than driver skill disabled the frontrunners.

Ill fortune can have a disproportionate effect in golf, too. Certain golf courses, such as Carnoustie in 1999, have so many hazards on them that even a highly skilled golfer has a very high chance of hitting a treacherous bunker or losing a ball in the rough. In these circumstances, a game of skill (akin to chess) is turned into a game of chance (more like snakes and ladders), and the more there is of the latter, the greater the opportunity there is for a relatively weak contestant to come through the field and win.

Why underdogs aren't always underdogs

Of course, it might just be that the so-called underdogs are no such thing. There is a lovely illustration of how we can be fooled into thinking that a team is an underdog when this is not the case.

Here is the argument. Since 10 is greater than 7 (symbol '>') and 7 > 3, it follows of course that 10 > 3. And in fact this can be generalised to the statement that if A > B and B > C, then A > C. This is known as *transitivity*.

We sometimes wrongly assume that transitivity applies to other situations. For example, if team A usually beats B, and B usually beats C, then surely A will usually beat C? Not so. To discover a situation that defies the norm, make four dice to represent four different football teams. On the six faces of the dice, put the numbers shown in the table below. For example, Liverpool's die must have four faces with number 4 on them, and two with 0s. Everton should have a 3 on every face.

TEAM	NUMBERS ON THE SIX FACES					
LIVERPOOL	4	4	4	4	0	0
ARSENAL	3	3	3	3	3	3
WEST HAM	6	6	2	2	2	2
NORWICH CITY	5	5	5	1	1	1

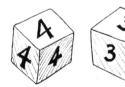

According to this rather artificial simulation, when Liverpool play Arsenal, there are only two possible scores: 4-3 to Liverpool, and 3-0 to Arsenal. If you play several dice games between these two teams, Liverpool will win about two-thirds of the matches.

When Arsenal plays West Ham, Arsenal wins 3-2 or Wet Ham wins 6-3, and in this case Arsenal wins about two-thirds of the matches.

West Ham-Norwich City has more possible results: 6-5, 6-1 or 2-1 to West Ham or 5-2 to Norwich City. Again, West Ham wins two-thirds of the games.

Liverpool usually beats Arsenal, Arsenal beats West Ham and West Ham beats Norwich City. So Liverpool will surely thrash Norwich City? Not in this game, they won't. Rolling Liverpool's dice against Norwich City, Norwich City amazingly wins two-thirds of the games.

This is an illustration of a non-transitive system. And, if it is possible here, perhaps something similar is possible when the real Liverpool and Norwich City play a football match.

Frequent changes of lead

One-sided contests are rarely memorable. There isn't much excitement in a rugby game in which England beats the USA by 50 points to nil. It's the close ones that we remember, especially if the lead changed hands towards the end of the match.

If one team is stronger than the other, then once it establishes a lead the chances are that it will hold on to it. But, if the competitors are roughly the same strength, then the frequency of the change of lead depends on the *effect of the lead* on the competitors' relative strength.

Lead changes are much more likely in some sports than in others. In some contests, being ahead influences the relative strengths of the competitors; in others it has no effect.

There are two well-known contests where establishing a lead actually has a reinforcing effect on the leader. One is the Oxford-versus-Cambridge boat race. In this case, once one boat is sufficiently far ahead of the other, the conditions actually favour the boat that is ahead, since it can occupy the middle water. Lead changes are almost unheard of. The Monaco Grand Prix is similar, because overtaking is so difficult. Even football can show these tendencies: if the away team scores a goal, it then plays defensively and the chance of any more goals in the match is reduced.

In other situations, however, establishing a lead probably has very little influence on what happens next. These situations can be analysed by what is known to statisticians as a *random walk* (see the box).

Some of the mathematics of random walks is complicated, but the conclusions make interesting reading. Once one side has established a lead, random-walk analysis says that the lead will change relatively rarely.

Another conclusion from random-walk theory is that, in an even contest, whoever is leading halfway through the game has a 50 per cent chance of continuing to lead throughout the rest of the game. Is this enough to make the second half exciting? Probably just enough. In the Oxford-Cambridge boat race, however, which is not a random walk, the leader at halfway almost always remains ahead thereafter, which is why it is normally such an anticlimax.

One way to increase the chance of a change of lead at the end of the game would be to make more points available in the later stages. If a goal counted as two points instead of one in the last ten minutes of a match, this would generate more occasions when there is a late switch of leader. Interestingly this is how some TV game-shows work. In Channel 4's *Countdown*, for example, most of the word rounds produce only 5 or 6 points

Random walks, deuce and snooker leads

Imagine a long straight road with a line down the middle. There is a hedge to the left of the road and a stream to the right. A drunk starts in the middle of the road, and attempts to walk along it. With each step he randomly lurches to the left or the right, with an equal chance of either direction. One question immediately comes to mind. How long before he falls into the hedge or the stream? It depends, of course, how many left or right steps it takes to reach the edge of the road. But, if the edge is N steps from the centre, then it turns out that on average he will take N^2 steps before he finishes his walk, though it could be fewer or more than this.

The drunk's progress is known as a random walk, and is the metaphor for a whole range of interesting problems in probability, including sport. The example above of falling into a hedge is directly analogous to a tennis game at deuce. In this case it takes two steps either way for the game to end (the equivalent of the drunk falling over), so on average – if each point has a 50-50 chance of going to either player – it takes 2^2 (i.e. 4) points for the game to end.

A second question of the drunk could be 'How often does he cross over the white line in the road?' This is the same as asking 'What is the chance in a snooker match that the lead swaps from one player to the other?' If the number of frames played in the match is F and the players are of equal strength, the answer turns out to be, on average, roughly $(\sqrt{F})/3$. So, after nine frames have been played, the lead is expected to have changed over only once. Over 100 frames, the average number of lead swaps is about three – fewer than intuition would suggest.

for a player, but the final round – the Conundrum (an anagram) – is worth 10 points. This gives the trailing player one final opportunity to overhaul their opponent.

Children's fiction has thrown up a more extreme example of how the points scored in the final stage of a game can have a disproportionately high value. In Harry Potter's wizard sport, Quidditch, goals count for 10 points but catching the 'snitch' at the end of the game scores 150 points. For goals to influence the result, one team therefore has to be at least 15 goals ahead, and none of the games in the early Potter books come anywhere close to this. Catching the snitch therefore counts for almost everything, though goal difference might come into play.

The disadvantage of increasing the points for later stages of the contest is that it rather negates the point of the early stages. The contest turns into a last-lap sprint with the rest of the contest becoming almost meaningless – the frenetic end to cycle track races after many laps of sedate cruising comes to mind. So far, sports administrators have held back from awarding more points for the final contests, but maybe the day will come.

The favourites meet for the decider

People may enjoy seeing the underdogs win, but the greatest sporting occasions are usually the battles between the giants, especially if the winner of these contests receives the ultimate trophy or gold medal.

The ideal type of tournament for building up to a big-match decider is the knockout. The FA Cup is one of the longest-standing examples of this. In the FA Cup, before each round the teams are drawn randomly from a bag. This means that there is a chance that the two 'favourites' might be drawn to play each other long before the final, thus knocking one of them out. Likewise, weak sides might be fortunate and end up being drawn against other weak sides, thus making far more progress in the tournament than they had any right to expect.

What the knockout cup doesn't do, therefore, is guarantee that the best two teams meet in the final. In fact, there is never more than a 2-in-3 chance that the best teams will meet in a Cup final – even if both of them get through to the semifinals.

Suppose that the favourites Rangers and Celtic are both through to the last four, along with two 'minnows', Falkirk and Alloa. The names are now drawn out of the hat. Here are the possible semifinals:

RANGERS V CELTIC	FALKIRK V ALLOA
OR RANGERS V FALKIRK	CELTIC V ALLOA
OR RANGERS V ALLOA	CELTIC V FALKIRK

All of these three sets of fixtures are equally likely, and in one of them Rangers and Celtic meet. So the chance that Rangers and Celtic will meet in the final is 2/3, assuming that both are near certain to beat the other teams.

At the start of the competition, it is even less likely that the 'big guns' are in different halves of the draw. If there are eight teams left in the contest, the chance of two chosen teams being in different halves of the draw is 4/7, or 57 per cent. If there are sixteen teams, the chance drops to 8/15. In fact, with N teams left, the probability of teams staying apart until the final turns out to be $N/(2N-1)$, which tends towards 50 per cent as N becomes large. So, in about half of all tournaments, Rangers and Celtic will meet in an earlier round than the final, assuming that one of them isn't knocked out first.

What this means is that knockout tournaments are a bit of a lottery, they don't necessarily reward the strongest teams. Some sports do something to correct this. The Wimbledon tennis tournament, for example, seeds the top players (32 of them at last count), and designs the draw so that these players cannot meet each other until the tournament has reached the last 32.

Furthermore, the top sixteen seeds are in separate parts of the draw until the last sixteen, the top eight seeds cannot meet until the last eight, right on to the top two seeds, who are in separate halves of the draw and cannot meet until the final. As a result, it is extremely rare for a non-seed to make it to a tennis final or even to the last four, in contrast with the FA Cup where non-Premiership sides have made the semifinals on a number of occasions.

The fairest way of identifying the best team is almost always a league, where each team has to play every other team at least once. The perfect climax to a league would be if the final match of the season was between the best two sides, but since the league fixtures are arranged before the season begins this rarely happens.

It's a knockout

Including qualifiers, 596 teams participate in the FA Cup; 282 men compete for the singles at Wimbledon. In both cases, the strongest participants don't enter until the later stages. With this limited information, how quickly can you work out how many matches there are in each tournament (ignoring any replays)?

The answer is surprisingly simple. The number of matches in a knockout tournament is always one smaller than the number of participants, making 595 matches in the FA Cup and 281 at Wimbledon. The reason is that each match knocks out exactly one participant, and at the end of the tournament only the winner has not been knocked out.

However, some sports have successfully combined the fairness of a league with the excitement of a knockout by having play-offs at the end of the season. Typically, the top few sides in the league go on to the knockout phase, which is set up so that the top two teams cannot meet until the end. This league/knockout combination is a fundamental part of not only the FIFA World Cup, but also many of the team sports in the USA. The authorities may not be able to guarantee a thrilling climax, but they know how to give themselves a great chance of one.

14

WHY DO KARAOKE SINGERS SOUND SO BAD?

How waves and fractions make good and bad harmony

The inventor of the karaoke machine has a lot to answer for. Before it existed, most amateur singers restricted their performances to the shower. Now, with microphone in hand and musical accompaniment for moral support, they can inflict pain on a far larger audience.

The reason why so much karaoke singing sounds so bad, of course, is that many people are unable to sing 'in tune'. Put another way, the sounds being emitted from the karaoke singer's voice box clash with the notes on the musical accompaniment – or with the notes that the audience's brains expect to hear.

There are several factors that determine whether singers sound out of tune. Some are to do with the expectations set within our culture, others are to do with the way the brain interprets sounds. But some of the reasons can be explained mathematically, and these are the main concern of this chapter.

It all begins with the curvaceous *sine wave* …

Sound waves

Ask anyone to think of a wave and they are most likely to picture a steady undulation, such as is found on the sea. The simplest such undulation is known as a sine wave, which looks like this:

The word 'sine', by the way, comes from the Latin word sinus meaning a bay, and sine curves look a little like bays found on a coast. Sine waves are in fact the most basic and fundamental waveform, and crop up in various real-world situations. For example, if you hang a weight from a spring, pull it downwards and release it, the weight will now bob up and down, and let's suppose it takes one second to return to where it started.

The distance of the weight from its central position plotted against time looks like this:

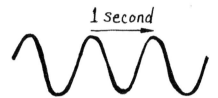

In this case, the time between the peaks of the wave, a *cycle*, is one second. The number of cycles per second is known as the *frequency* of the wave, so in this case the frequency is 1 cycle per second, or 1 Hertz (often written as 'Hz').

Sine waves can also be created by motion in a circle. If you took a ride on the London Eye and plotted your height above the ground against time, the graph would look like this:

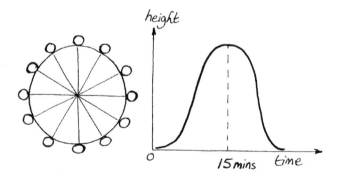

The position of a capsule on the wheel over time plots out a sine wave, and, since a complete ride on the wheel takes about half an hour, the frequency of the wave is 1 cycle of the wheel per 30 minutes, which is 1 cycle per 1,800 seconds or about 0.0006 Hz.

Anything that vibrates or cycles sends pulses through the air, causing the air molecules to move to and fro like the weight on a spring. The pulses are sound waves, and the human ear can detect these sounds as long as they are between about 20 Hz (a very low note) and 20,000 Hz (a high whistle). The frequencies of the Millennium Wheel and the weight bobbing on the spring are too low to be picked up by our ears, but if the spring was more powerful or the Wheel was whizzing round at a sickening speed, both would create audible notes. The sound would resemble that made by a tuning fork, or the penetrating noise of a wet finger running around the rim of a fine wineglass.

Other vibrating objects such as bees' wings, struck saucepans and electric razors all make notes, too. The greater the frequency of the vibration, the higher the note, and, since notes put together make tunes, with the right collection of bees, saucepans and electric razors, you could put on a recognisable, if rather eccentric, performance of Beethoven's Fifth Symphony. Or any other tune, of course.

The waveforms of these unconventional musical instruments are complicated, but it turns out that the sine wave is fundamental to them all. A Frenchman called Fourier made the remarkable discovery that absolutely any wave, however irregular it looks, can be constructed simply by adding together a combination of different sine waves. It might, for instance, look like this:

The difficult part is working out exactly *which* sine waves you need to add together. The squiggle above may be a combination of ten or more waves, each with a different frequency and amplitude, and the analysis required to find them is well beyond this book – though not necessarily beyond our ears.

How the ear reacts to combinations of notes

Without knowing anything about Fourier the human ear is able to listen to a sound wave and to some extent break it down into its constituent sine waves. If you listen to notes being played on three recorders simultaneously, you will probably be able to pick out three distinct notes, even though an oscilloscope monitoring the sound would show that a complicated-looking combined wave is reaching your ears.

However, although ears are good at picking out note combinations, they are not perfect at it. If there are two pure notes of identical frequency, the human ear will detect a single sound. The ear can only detect different notes if the frequencies are sufficiently far apart. Here is a rough guide to what happens, though the exact frequency ranges involved depend both on the individual (some people have more acute hearing than others) and on the frequency level being listened to.

If the difference in frequency is tiny, less than 1 Hz, say, only one note is heard, and the ear is quite happy. Two violins in a professional orchestra cannot ever be played at exactly the same note, but they are so close that few humans can tell the difference.

If the notes differ by between 1 and 10 Hz, the joint sound that the ear detects will be a single note that pulses loud and soft, a phenomenon known as *beats*.

If the notes differ by between 10 and 20 Hz, there is a rough sound, caused partly by high-frequency beats. The ear doesn't like this at all. In fact it is believed that this rough sound caused by frequencies that are within a certain critical range of each other, is the basis of what sounds universally 'bad', or dissonant, in music of all cultures.

When the frequencies differ by more than a critical range – say 20 Hz – then the ear can clearly identify that they are different, and regards their combination as being quite acceptable, though not necessarily beautiful.

This simple theory suggests that, as long as their frequencies are far enough apart, two pure notes played together should always sound OK.

Does this mean that a bad karaoke singer is somehow managing to be out of tune just enough – but not too much – to be producing frequencies that perfectly clash in the critical band of the backing sound? Partly, yes. However, the notes from a singer's larynx are not pure sine waves. They are a combination of many different frequencies, and even if they are far apart, these impure notes can produce nasty clashes …

Why do two different notes sometimes clash?

When any instrument is plucked or struck or blown, it will produce a note at its natural frequency. However, instruments will produce notes at other frequencies at the same time. These other frequencies are known as *harmonics,* and on well-crafted musical instruments such as flutes and guitars (but not razors and saucepans) these harmonics are simple multiples of the basic frequency. So, if a plucked string has a base frequency of 100 Hz, it will also produce quieter frequencies at multiples of 100 Hz:

BASE NOTE	1ST HARMONIC	2ND HARMONIC	3RD HARMONIC	4TH HARMONIC	etc
100 Hz	200 Hz	300 Hz	400 Hz	500 Hz	...

The base note and the harmonics are all pure sine waves, but their combination is a much more complex-looking waveform, found by adding the harmonics together. These harmonics make the note played on a piano, for example, sound much 'richer' in quality than the sound produced by the finger around the wineglass. If you strike a single piano key very hard, then in addition to the main note you may well be able to detect these higher notes in the background. In flutes, the base note is much more dominant, with very little in the way of audible harmonics. Whatever the instrument, you are unlikely to be able to hear anything beyond the fourth harmonic.

So what about listening to two notes that are played at the same time? What you hear is a combination of their lowest frequencies and all of the harmonics. This combination can sound good, especially if the relative string or tube lengths of the notes are ratios of small whole numbers, like 2/1 or 3/2. To see why, here are the harmonics of three strings, one full-length, one half-length and the other two-thirds-length.

The longest string creates the following harmonics:

BASE NOTE	1ST HARMONIC	2ND HARMONIC	3RD HARMONIC	4TH HARMONIC	5TH HARMONIC
100 Hz	200 Hz	300 Hz	400 Hz	500 Hz	**600 Hz**

Adding sounds together

When two sound waves are produced at the same time, the waves can be literally added together. Two identical pure notes added together make a note of identical frequency – but louder!

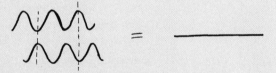

The timing of the peaks and troughs of the above notes are identical, or 'in phase'. Even if they are not in phase, two identical sine waves added together always combine to create another sine wave of the same frequency.

If the peaks of one coincide with the troughs of the other, they completely cancel each other out when added together:

These two notes combined would actually appear silent to the listener. This principle is used by engineers to create anti-noise – noise generators that produce identical but 'upside-down' sound waves in order to reduce the noise level in the environment.

The frequencies of the half-length string are all double those of the longer string:

BASE NOTE	1ST HARMONIC	2ND HARMONIC	3RD HARMONIC	4TH HARMONIC	5TH HARMONIC
200 Hz	400 Hz	600 Hz	800 Hz	1000 Hz	1200 Hz

And the frequencies of the two-thirds-length string are in between those of the other two strings:

BASE NOTE	1ST HARMONIC	2ND HARMONIC	3RD HARMONIC	4TH HARMONIC	5TH HARMONIC
150 Hz	300 Hz	450 Hz	600 Hz	750 Hz	900 Hz

When plucked together, any pair of these notes has coinciding harmonics. Indeed, the 600 Hz harmonic is shared by all three strings. The ear, which likes coinciding frequencies, will appreciate this. Furthermore, none of the frequencies are close to any of the others, and this means there won't be any of the harsh beating noises that the ear dislikes.

These three ratios of string create the most comfortable harmonies of all, which helps to explain why notes in the ratio of 3/2 and 2/1 have appeared in music of almost all cultures, past and present. Even an ancient Chinese flute discovered by archaeologists was found to have holes positioned to produce 3/2 notes. This ratio is known in musical parlance as a *perfect fifth*.

The general principle is that notes that sound good together have some harmonics that coincide with each other, and have no harmonics that enter each other's critical 'bad noise' frequency band. By far the best such harmonies are produced by the ratios of small numbers, such as 3/2, 4/3 and 5/3.

Pythagoras and the clanging hammers

The story goes that Pythagoras was one day walking past a forge when he heard two hammers banging together. The notes that the two hammers produced sounded the same – and yet different.

When Pythagoras checked, he discovered that one piece of metal being struck was exactly half the length of the other, and the shorter piece was producing the higher note. What he had heard would later become known as an octave. Pythagoras was able to reproduce this effect by plucking strings of different lengths. He went on to experiment with other simple ratios of string lengths. Notes whose lengths had simple ratios such as 3/2 or 4/3 seemed to sound good together or to 'agree', the Greek word for agreement being 'harmonia', hence harmony. All of this reinforced Pythagoras's notion that numbers were behind everything in the natural world.

Why there are twelve different notes

Music has far more notes than just the simple octave and the 3/2 note. In fact there are twelve notes in a Western octave, and most people give no thought to where these notes come from. They just seem to be 'there' in much the same way as snowflakes always have six points. However, this system of twelve notes evolved through a combination of mathematics and chance. Since this music scale is partly what we use to judge a karaoke singer, it's worth knowing its origins.

Pythagoras was the first person in Western culture to produce a musical scale. He decided that the scale should contain exactly seven different notes, partly because of the mystical importance of the number seven. He also thought that all notes should be constructed using ratios of 3/2.

The first music scale?

This is what the original Pythagorean music scale might have looked like. Each string length has been created by multiplying 2/3 or 3/2 together. To bring all the notes between 2 and 1, a single octave, their lengths have then been doubled or halved (doubling or halving a string length changes a note by an octave but retains its essential sound.) So, for example, 2/3 x 2/3 = 4/9 or 0.444. To bring this between 2 and 1, it needs to be doubled twice, making it 16/9 or 1.778. In the scale that follows, the longest string makes the lowest note:

NOTE	MODERN NAME OF NEAREST NOTE *	RELATIVE LENGTH OF STRING TO MAKE NOTE
1st (base note)	D	2/1
2nd	E	16/9
3rd	F	27/16
4th	G	3/2
5th	A	4/3
6th	B	32/27
7th	C	9/8
8th (octave)	D	1/1

In other words, if you were to play the white notes on a piano in this order, the scale would sound very similar to the Pythagoras scale. Incidentally, the seven notes didn't acquire letters until medieval times. When playing a complete scale, the first note is repeated at the end to make the octave.

As it turns out, the scale attributed to Pythagoras produced reasonably good harmonies, but these were not necessarily any easier on the ear than those being created independently in other cultures. Some cultures settled on a 5-note ('pentatonic') scale, for example, while others produced scales with as many as 22 notes.

If you were to listen to this Pythagorean scale being played, the notes would sound pretty close to those that we are familiar with in modern music. However, medieval musicians realised there was something missing. They wanted the freedom to start a tune on any note and still be able to sing along to a familiar melody. This was not possible using the seven notes of Pythagoras's scale.

Try playing 'Three Blind Mice' on Pythagoras's scale. If the first three notes played are 'E, D, C' it sounds all right, but if you attempt to play 'Three Blind Mice' starting 'F, E, D' it sounds quite wrong. Why? Because the notes in Pythagoras's scale were not spaced evenly apart.

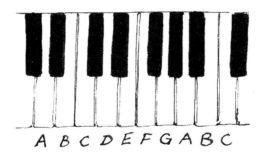

Medieval musicians had to insert notes to give the scale a roughly even spacing, represented today by the black notes on a piano keyboard. One way to fill in the gaps was to extend Pythagoras's idea of creating all of the notes using the 3/2

interval. As it happens, if you start with a string that produces a note and reduce its length by the ratio two-thirds twelve times, you end up producing a note that is almost exactly the same as your starting note, but seven octaves higher. This is because $(3/2)^{12}$, which works out as 129.7, is approximately 2^7, or 128. The coincidence of 129.7 being so close to 128 is the main reason why there are twelve notes in the modern scale. Do the numbers seven and twelve ring a bell? For quite different reasons, they are also the basis of the western system of measuring time, as discussed in Chapter 1.

The twelve notes created by 3/2 intervals make a reasonable scale, though some of the intervals between the notes sound pretty awful. To improve the harmonies between different pairs of notes, the Medievalists began to experiment with different

---

Wolf fifths and the devil's interval

Musicians have had to grapple with some unpleasant sounds over the centuries, all of them caused by note frequencies that weren't in simple proportions. In the Renaissance musical scale, the toe-curling sound made when B and F were played together was a particular nuisance. Surely God had never intended such a noise to be heard. It was nicknamed the devil's interval, and for a while the church banned it from all music. In later scales based on Pythagorean ratios, the so-called 'fifth' formed by playing (say) F-sharp and D-flat together, made a discordant sound reminiscent of a howling wolf, earning it the name a 'wolf fifth'.

ratios of string lengths. Pythagoras's idea that every note had to come from the ratio 3/2 was sensibly dropped – why not include the ratios 5/4 and 5/3 too? Later inventors of the music scale struggled to find string lengths that produced simple ratios (and therefore pleasant harmonies) for all twelve notes when plucked together. But perfect combinations for every pair of notes proved elusive.

There is one way to remove the occasional cringe-making harmonies from a musical scale, and that is to make the intervals between every one of the twelve notes identical. Eventually, somebody spotted that the way to do this was to make the lengths increase in a *logarithmic scale*. What this means in practice is that each note's frequency has to be 1.06 times higher than its predecessor (1.06 is approximately the twelfth root of 2). As a result, on a modern keyboard you can play 'Auld Lang Syne' or 'Happy Birthday' starting at C, E, F# or anywhere else and it will always sound like the regular tune.

Back to the karaoke singer

All of which brings us back to the failure of the karaoke singer to hit the right notes. What he is failing to hit, of course, are the notes to which we have become accustomed in our Western scale. The sound waves he produces include frequencies that grate with us or make us want to howl like wolves.

It isn't all down to maths, however. We shouldn't forget the cultural factor. Some of the notes and harmonies were established because they sound good to most people, but others were created because the mathematics of the different ratios dictated that they had to be there to fit into the musical scale. Some of these latter notes arguably only sound good because we are so used to hearing them.

So, although the karaoke singer may sound bad, it isn't entirely his fault. Although some intervals seem to be popular in just about all cultures – the fifth, for example – others that we are used to are very specific to the West. Other cultures have their own completely different scales, created from different, less mathematical starting points. Perhaps there is an island somewhere in the South Pacific where even the most terrible of karaoke singers sounds as mellifluous as the birds, and as perfect as Pavarotti does to us.

15

HOW CAN I BE *SURE?*

The art of proving things

Young children love colouring-in pictures, and given the chance most will try to use as many different crayons as possible. Of course one rule of colouring-in that any child understands is that neighbouring areas must always be a different shade. So, what if the challenge is to keep the number of crayons to a minimum? How many different crayons are needed to be sure that no two neighbouring areas in any picture are the same colour?

After a little experimenting, it's easy to find situations where at least four different colours are needed. Here, for example, is a map of part of Europe, including a bit of sea.

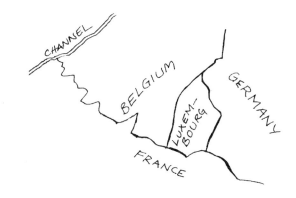

Whatever combination you try, Belgium, Germany, France and Luxembourg must all be different colours, because all are touching each other. The English Channel, on the other hand, can be the same colour as Luxembourg because the two do not touch.

For the four-colour rule to apply, 'neighbouring' means sharing a border, not meeting at a point. For example, here is part of the USA:

If meeting at a corner did count as neighbouring, all four states would have to be different colours and a region surrounding all four would then have to be a fifth colour. But, if only *shared* borders count, then once again four colours will suffice for the entire USA.

Needing at most four colours on a map certainly *seems* to be the general rule, but is it true for all possible cases? In 1852, a student called Francis Guthrie made this very conjecture. It seemed a simple enough problem, yet the *proof* of it turned out to be incredibly difficult. Francis asked his brother Frederick, who didn't know the answer, and Frederick asked Augustus DeMorgan, one of his lecturers and a great mathematician of the time. DeMorgan didn't know either, and this simple query turned into one of the most popular maths challenges of the next century, namely to prove that all maps can be coloured-in using at most four colours.

After thousands of attempts, nobody managed to find an example that required five colours. All of this was pretty convincing evidence, but it did not, in mathematical terms, constitute a proof. There was always the possibility that some obscure exception existed that had not been discovered.

In fact, although progress was made in proving the conjecture, it took the arrival of computers to enable the final step to be taken. In 1976, with the aid of hundreds of hours of computer processing time, the mathematicians Appel and Haken finally turned the four-colour conjecture into the *four-colour theorem*. And mapmakers who had known it for years said, 'Told you so!'

All of this highlights an important part of maths that hasn't really come to light elsewhere in this book. Proof lies at the very heart of mathematics. Using rigorous logic that builds from

---

Precision versus approximation

How can you tell the difference between a mathematician and an engineer? Ask them what pi is.

Mathematician: *'It is a ratio describing the circumference of a circle to its diameter, a transcendental number which begins 3.14 and continues for an infinite number of digits.'*

Engineer: *'It's about 3, but let's call it 10 just to be on the safe side.'*

---

certain accepted truths, maths can prove all sorts of abstract, and not-so-abstract notions. Mathematical proof is 100 per cent, concrete certainty.

As far as everyday life is concerned, however, mathematical proof is not so obviously relevant. In the four-colour theorem for example, hundreds of trials without finding a contradiction is perfectly good enough 'proof for most people, especially since much of the time we have to make decisions in life on just a single observation.

In fact even to mathematicians proof is often little more than what somebody once called a 'hygiene' factor. To answer a question, mathematicians often use imagination and guesswork first, and only then does proof come in to play to ensure there hasn't been a grotesque error somewhere in the workings.

Despite this, proofs do have some everyday relevance. The sort of thinking that mathematicians use in abstract problems is a good discipline to use informally in solving more everyday issues. How, then, do mathematicians go about tackling problems and proving their answers?

Odd socks, and proof by examining every outcome

Perhaps the most laborious type of proof is the one where every possible outcome is examined. For example, how do you prove that the word 'xylophone' did not appear in the previous chapter? There is little choice but to check through each word or, to save some effort, check every initial letter of every word. No x's means no xylophone.

This type of proof can be used for an old sock problem, too. To get around the problem of lots of odd socks, one of the authors has adopted a simple strategy. All he ever buys are identical black socks and identical blue socks. That way, he always has lots of pairs, even if he loses the occasional sock. But, on dark

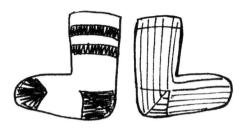

winter mornings, black and blue socks look remarkably alike. If he has ten black and ten blue socks in a drawer, how many socks does he have to take out to be absolutely certain that he has a pair of socks in his hand?

To some people, the answer is obvious. Only three socks need to be taken from the drawer. However, others argue that, to be *certain* of ending up with a pair, eleven or even nineteen socks have to be taken out.

One way to prove how many socks are needed for a pair would be to look at every single possible order in which socks can be removed from the drawer: black, black, blue, black; blue, black, black, blue … and so on. This would, however, take rather a long time, since there are nearly 200,000 different combinations. A simpler proof requires far fewer combinations, by using a logical short cut. Imagine taking any two socks from the drawer. If they make a pair, then the problem is solved in just two socks. If they don't make a pair, they must be odd, i.e. one black and one blue. Since the next sock out of the drawer is going to be one of these two colours, the three socks must certainly contain a pair. Hence three socks is the maximum needed.

Surprisingly, the sock problem has a lot in common with the four-colour theorem. An exhaustive checking of a huge number of options finally proved the latter, but, as with the odd sock, there were short cuts to reduce the range of possibilities that needed to be tested.

Exhaustive testing is all right as a way of finding a proof, but if it looks as if it is going to take a long time, it's worth thinking about how the search could be simplified. And, besides, short cuts are more satisfying.

The pigeonhole proof

There is one type of short cut that carries the name pigeonhole proof. At the peak of the National Lottery's popularity, over 15 million people bought tickets in a single week, and somebody in the press wondered if everyone had picked a different combination. There are, of course, millions of different combinations of numbers that could be chosen, but how can we be sure whether 15 million people chose different combinations or not?

Here is how. The total number of different combinations that you can choose for a lottery ticket is 13,983,816. Let's take the extreme case. Suppose that, for the first 13,983,816 tickets sold, every person chose a different combination. Every single possible combination had now been used, and without a single duplicate. What about the 13,983,817th person? Since every combination had already been entered, he had no choice but to choose a combination that somebody else had already chosen. Hence it is *certain* that, if 15 million people buy lottery tickets, at least two of them choose the same numbers. Of course we don't know what that coincidental combination is, but that isn't what is being proved. All that has been proved to be universally true is that there was at least one combination that appeared at least twice.

This is a pigeonhole proof. You can imagine creating a pigeonhole for every possible choice of lottery numbers. You then try to place each selection into a different pigeonhole.

When you run out of empty pigeonholes you have no choice but to put a second entry into one of them.

Exactly this principle can be used to prove that, in Manchester United's last home game, there were certainly at least two people who were born on the same day of the same year. How can we know this? Let's guess conservatively that there were only 50,000 at the game (there were probably more like 70,000), and that everyone in the crowd was aged between 0 and 100 (another very conservative estimate – the age range was probably far narrower than that). All of those people were born in the last 100 years. Could they all have been born on different days?

In each year there are 365 days, plus the odd leap year, so in 100 years there are no more than 36,525 different birth dates. So, by the pigeonhole principle, if a crowd is bigger than 36,525 people then there is certain to be a duplicate. We can say this with absolute confidence, even though we've no idea if the coincidental date is 13 July 1961 or 22 September 1974 or any other specific date.

You can probably invent other pigeonhole proofs of your own for all sorts of trivia. How many people do you need before you can be certain that two of them have the same number of hairs on their head? How many books would it take before you could guarantee that two of them had exactly the same number of words?

Proof by contradiction

Imagination can play an important part both in problem solving and in finding proofs. Often, for example, it helps to prove a point by first of all imagining something ridiculous, for example that the complete opposite is true. As a trivial example, there are lots of ways of multiplying two positive numbers together to get the answer 72 (for example 2 x 36, 5 x 14.4). How might a mathematician prove that at least one of the two numbers in this multiplication must be greater than 8? One way is to say, 'Suppose *neither* of the numbers is bigger than 8.' What would be the consequences of this?

We know 8 x 8 = 64, which is smaller than the 72 we seek. What if one of the numbers is less than 8? 'Less than 8' x 8 is less than 64. And what if *both* are less than 8? 'Less than 8' x 'less than 8' is also less than 64. So we have demonstrated that the numbers cannot both be 8, and nor can they both be smaller than 8, so we have found a contradiction. Therefore, at least one of the numbers must be more than 8.

All right, that was not an earth-shattering example, but what matters here is the principle of imagining an answer and seeing where it takes us.

This approach of making an assumption and then following it through until it leads to a conclusion that we know is false carries the formal Latin name of *reductio ad absurdum,* literally 'reduction to the absurd'.

Informally, this technique is used in conversation all the time. Barristers and politicians are particularly fond of it as a means of demonstrating the weakness of an opponent's argument. Hansard, the written record of parliamentary debates, is no doubt full of arguments along the lines of: 'The Right Honourable Gentleman claims that he will increase public spending. The only way in which he can achieve this is by increasing taxes – which he has already ruled out. I therefore pronounce his argument to be in tatters.'

Perhaps the other most common everyday application of this method of proof is in puzzle solving. Newsagents sell thousands of books of 'logic puzzles' every week, which means that tens of thousands of hours are being spent unravelling problems such as this:

Three women live alone in a row of three isolated houses on a hill. Here are three facts that refer only to them. (1) Maureen does not live in the middle house. (2) Debbie shares her lawnmower with the architect. (3) Jan lives two doors up from the artist. Who lives where?

The standard way into these puzzles is to make a guess and test it out. For example, let's suppose that Debbie is the artist. From fact (3), this means Debbie must live two doors from Jan, which we can represent like this:

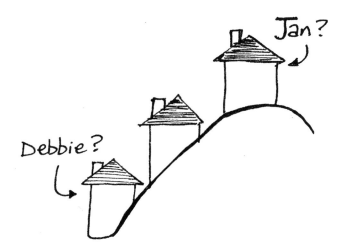

However, this would mean that Maureen occupies the middle house. This is a contradiction of fact (1) which says she *doesn't* live there. Hence our initial assumption is false, so Debbie is *not* the artist. From here it can quickly be concluded that Debbie must live in the middle house, Jan in the top house and Maureen in the bottom one. Phew, that's over.

More formally, *reductio ad absurdum* has been used in mathematics for centuries. One of the most famous examples is Euclid's proof that the square root of two is not a ratio of two whole numbers. His approach is built from the same sort of initial supposition as the one about Debbie, in other words, 'Let's start by supposing something, and see if it leads us to a contradiction.' In Euclid's case he starts by supposing that the square root of two *is* the ratio of two whole numbers, and, sure enough, it all ends in tears. (The simple proof is in the Appendix on page 217.)

Starting small and working upwards

Another helpful approach in tackling life's problems is to start by thinking about the question in its simplest form, and work up from there. After all, a lot of problem-solving goes wrong simply because the issues are too complicated to take in all at once.

A similar principle of starting with simple examples often helps in mathematical problem solving too. And it can help in some proofs. Here is a nice example. Everyone knows that $3^2 - 2^2 = 9 - 4 = 5$. But what is the answer to this sum:

$$222{,}222{,}222{,}222{,}222{,}222{,}222^2 - 222{,}222{,}222{,}222{,}222{,}222{,}221^2$$

Most calculators won't help here, since they won't handle numbers this big. And even desktop computers are likely to get it wrong. One computer gave the answer zero, which is patently nonsense since the difference between these two square numbers must be a huge number itself.

One way to tackle the problem is to start small and look for patterns.

$$1^2 - 0^2 = 1$$
$$2^2 - 1^2 = 3$$
$$3^2 - 2^2 = 5$$
$$4^2 - 3^2 = 7$$

There seems to be a pattern here. To find the difference between two adjacent squares, it looks like all you have to do is add the unsquared numbers together. For example $2+1 = 3, 4 + 3 = 7$, and so on. However, this idea is still only a hunch. How can we be sure that this pattern goes on for ever?

One way is to draw pictures using dots. Here are the first four squares.

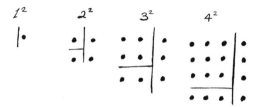

Do you see how each square can be made from the predecessor? To get from 2 x 2 to 3 x 3, all you need to do is add 2 to one side of the square, then 3 to the other side. To get from 3 x 3 to 4 x 4 add 3 to one side and 4 to the other. Put this way, it is obvious why the difference between two adjacent squares will always be the smaller + the smaller plus one, and also that it will *always* work from one square to the next.

This, rather informally, is a proof. In fact, since it shows how the theory is true for the simplest case and then shows how the series must continue on from there, this is known as a proof by induction.

So what is the answer to that brute of a sum earlier that defeated even a computer? Easy!

It is 222,222,222,222,222,222,221 + 222,222,222,222,222, 222,222, which is 444,444,444,444,444,444,443.

Proof by pictures

In that previous example, an illustration was used to prove the rule. Pictures are often a far more powerful way of proving things than more abstract-looking algebra.

One wonderful example is the proof of the 'chessboard problem'. A chessboard is a grid of 8 x 8 squares, or 64 squares in total. Suppose you have cut off two opposite black corners of a chessboard, like this:

Now you are given 31 dominoes, each of which is exactly the size of two chess squares stuck together. Altogether, these dominoes will cover 62 squares on a chessboard, and as luck would have it, that is exactly the same number of squares that are left on the board above.

The question is: can you find a way of covering all the squares on the board using the 31 dominoes?

You would think this might be quite easy, but, after several attempts, it soon becomes clear that it is not such a simple challenge after all. When you are down to the final domino, the two remaining uncovered squares never seem to be next to each other. Maybe this problem can't be solved after all.

Coconut picking, the mathematical way

A sailor and a mathematician were trapped on a desert island, on which there was a tall coconut tree with just two coconuts on it. The men had no food, but they didn't fancy climbing up for the coconuts because both of them hated heights. Finally hunger got the better of them. They tossed a coin, the sailor lost, and, sweat dripping from his brow, he proceeded to shin up the tree, lean tentatively across, loosen one coconut and watch it fall to the ground. 'OK, I've got to come down now,' he said, shinning down, and the two shared their spoils. The next day they were hungry again. 'Your turn', said the sailor. He then watched with some puzzlement as the mathematician pieced together the shell of the eaten coconut, stuck it together with seaweed, and then proceeded to shin up the tree with the coconut under his arm. Hanging precariously, the mathematician then reached out and hung the old coconut back on the branch it had come from before easing himself back to the trunk and shinning down. 'What are you playing at?'screamed the sailor. 'Well, now we've reduced the problem to one that we've already solved', said the mathematician.

There is a clever proof that will settle the matter. Look at the chessboard with the missing corners. Both of the removed pieces of the board were black. This means that the pattern you want to cover is made up of 32 white squares and 30 black squares.

Now imagine placing a domino over any two squares on the board. However you do it, you always cover one black square and one white. So, if you put down 30 dominoes, you will have covered all 30 of the black squares on the pattern, and 30 of the 32 white squares. You have one domino left, and the two uncovered squares are both white. Take a look at all the white squares on the board. They are never adjacent to each other, they are always on diagonals. Since two white squares are never adjacent, it will never be possible to cover the final two squares with your domino.

Chess expert William Hartston posed an interesting question about this problem. If this proof had not been available, how else could the problem have been proved? Despite the simplicity of the problem, it might be that any other proof would have had to be extremely long and convoluted. In fact, it seems as though it is almost a matter of luck whether a simple-looking problem has a simple proof, a complicated proof, or, in some cases, no proof at all.

If maths had been just a little different, maybe the four-colour theorem could have been proved in five minutes while the chessboard problem would still be baffling the top minds to this day. It seems that we can never be certain even about how to be certain.

The most proven theorem of all?

One of the most famous theorems of all is that of Pythagoras. Over 300 different proofs have been found for this theorem. Talk about overkill! As a reminder, the theorem says that the square of the two shorter sides of a right-angled triangle add up to the square of the longest side, or hypotenuse. For example, in this triangle, with two sides of length 3 and 4 ...

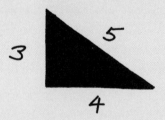

... the square of the hypotenuse will be $3^2 + 4^2 = 25$, which means the hypotenuse must be 5 units long. This 3, 4, 5 triangle is popular with builders, by the way, who like to use it to create right-angles for checking that corners are square.

16

CAN I TRUST WHAT I READ IN THE PAPERS?

How spin doctors conjure with figures

This final chapter is about magic. Things will appear from nowhere, or with a touch will suddenly grow ten times as big. You will see something become smaller and yet bigger at the same time, and all before your very own eyes. This is not the work of the witch doctor, but something far more sinister. This is the magic of the spin doctor, and his props are numbers.

Spin doctors help all sorts of people to manipulate the truth, but most often they are associated with politicians. There is nothing new about this. In the nineteenth century, Benjamin Disraeli declared, 'There are lies, damned lies and statistics', and no doubt propagandists had been using numbers to distort the facts before then, too.

This old craft acquired its new name of spin-doctoring from baseball, where a pitcher can deceive the batter by spinning the ball to make it swerve. The media seized on this metaphor to describe the way in which press and public-relations officers manipulate information in order to deceive the recipient in some way.

The aim of such spin is usually to make information sound better than it actually is. Numbers play a crucial role in this, taking advantage of the public's general discomfort with maths and their consequent reluctance to challenge the figures. It turns out, too, that numbers can be surprisingly helpful as a flexible tool in helping you to say what you want to say.

Making something out of nothing

Here is one of the simplest tricks of all. To illustrate it, presented below are two simple facts about a company called Cuddle-Co, who make a range of toy rabbits called Mr Snuggles:

LAST YEAR'S SALES
£ 500,000

THIS YEAR'S SALES
£ 515,000

Is this good news for Cuddle-Co? Of course it is, say the PR department, who are the in-house spin doctors. RECORD SALES FOR MR SNUGGLES! announces the headline. And this is quite true, the money earned by Mr Snuggles bunnies has never been so high.

So where is the trick? What PR have conveniently ignored, because it doesn't help their case, is that this year, as every year, there has been inflation, which just happens to have been 3 per cent. Every economy experiences an annual inflation rate, in which prices and wages increase. If prices go up by 3 per cent and wages go up by 3 per cent, then nothing has changed – every consumer has exactly the same purchasing power as in the previous year. Cuddle-Co's sales have gone up by £15,000/£500,000, which is precisely 3 per cent. In other words, nothing has changed in the business. 'No news' has been turned magically into 'good news'.

Conveniently ignoring inflation is probably one of the most common sleights of hand practised by spin doctors, and passed on to the public without challenge by the media. Everybody likes to see teachers' salaries, money spent on hospitals and the value of possessions rise every year, and because of the normal process of inflation they usually do. It all sounds like good news, but in itself these increases are meaningless. 'More' does not necessarily mean 'better'. Nor, of course, does it necessarily mean worse. By exactly the same argument, electricity bills, beer prices and the amount of tax raised by the government are all likely to rise each year (each one a 'shock-horror' story) and yet, because of pay rises, these increases may have no effect on people's standard of living.

Double-counting, or turning one into two

One of the classic conjuring tricks, of course, is making things appear from nowhere. A well-documented example of making money appear from nowhere was the so-called double-counting escapade of the government in 1998, much criticised in the press at the time. It was an early example of what the press has referred to ever since as *spin*.

Early that year, the Education Secretary, who was then David Blunkett, announced a £19 billion increase in spending on schools. Given that the total amount being spent per year at the time was £38 billion, this sounded like a quite astonishing investment – an increase of 50 per cent. This was great news for schools, and a big vote winner.

The £19 billion increase was statistically correct, but, as with most magic tricks, all was not as it might appear. To see why, consider a different example. Suppose that your local water company announce that, due to rising costs, they are going to have to increase your annual bill by £5 for the next three years.

CURRENT BILL	YEAR 2	YEAR 3	YEAR 4
£60	£65	£70	£75

This is clearly not good news, but just how bad is it? If you wanted to be really pessimistic you would say that water bills are going to go up by £15 – that's £15/£60, which is a 25 per cent increase.

However, that would be a little harsh. The £15 increase is still three years away. We've already seen that inflation has to be taken into account. A fairer statement by a neutral observer would be that, if annual inflation is about 3 per cent, water bills are going to increase not by 25 per cent but by about 15 per cent in *real terms* – that is, once inflation has been taken into account. In today's money, bills will have increased by about £10 and not £15. Bad, but not quite as bad as it first seemed.

How would you react, then, if a spin doctor announced to you that you are not going to have to pay £10 more, not even £15. In fact you face a shocking increase of £30 for your water, which is 50 per cent of what you pay at the moment? You are probably a bit surprised at this. Where has this massive bill come from? Is the water company hiding something?

Not at all. It's just a question of how you interpret the figures. In Year 2 you will pay £5 more than you currently pay; in year 3, £10 more; and in year 4, £15 more: £5 + £10 + £15 = £30! Strictly speaking this is correct, but it is certainly an unconventional way of using the figures. In fact, it probably reminds you of the missing-pound trick in Chapter 2.

Yet this is exactly how the new spending on education was presented. Education spending was set to increase as follows:

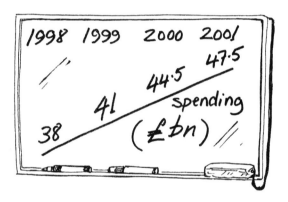

What was the increase in spending after 1998 going to be? By 2001, it was due to reach £47.5 billion, £9.5 billion more than in 1998. If you add up each lump of 'extra spending' for the three years, however, it adds to £19 billion. So what was the extra investment in schools, £19 billion, £9.5 billion, or taking account of inflation, less than £9.5 billion? How long is a piece of string?*

* Have politicians cleaned up their act? Of course not. Double-counting money remains a staple trick of every government. In 2018, Conservative Chancellor Philip Hammond announced that the NHS would receive a funding boost of £84 billion – impressive, when the current budget was about £115bn. But the £84bn was reached by adding up extra spending over several years, in a similar sleight of hand to David Blunkett's twenty years earlier.

Making something smaller and bigger at the same time

Percentages are a particularly good prop for performing the spin doctor's magic. Take, for example, the miracle of vanishing exports.

'I won't deny that these have been tough times for the company,' said the spokesman. 'Last year, due to the strength of the currency, our exports fell by 40 per cent, but I'm delighted to announce that, thanks to the outstanding efforts of our marketing team, this year has seen us bounce back with a sensational 50 per cent increase.' The shareholders are impressed – 40 per cent down followed by 50 per cent up. Sounds like a net increase of 10 per cent.

This is another classic piece of misdirection by the number magicians. Here are the real figures:

TWO YEARS AGO	100,000 units exported
LAST YEAR	60,000 units exported

So last year, the exports dropped by 40,000 from their previous 100,000 level. That is indeed a 40 per cent decrease. This year, we are told, has seen a 50 per cent increase over last year. Last year's exports were 60,000, and 50 per cent of this is 30,000, so after the 50 per cent increase we now have:

THIS YEAR	90,000 units exported

Hang on a second, that isn't a net increase of 10 per cent over two years ago. A drop of 40 per cent followed by an increase of 50 per cent led to a *net decrease* of 10 per cent. Incredible! How's it done? There's nothing hidden, that's just the way that percentages work. The spokesman compared the 40 per cent and the 50 per cent as if they were the same thing, but, since they were based on different starting figures, it was like comparing apples with pears.

Want to feed the five thousand? How to turn 1% into 50%

First spin doctor: *'Last year, the price of coffee went up by only two per cent. This year it has gone up by three per cent – that's an increase of only one per cent, which is quite reasonable given the poor crop this year'*

Second spin doctor: *'Not at all. If it went up by two per cent last year, and three per cent this year, that means it has gone up by fifty per cent!'*

1% or 50%? You choose.

Use averages to make everyone feel better – or worse

You can pull off a lot of tricks with averages. The whole concept of what 'average' means is a slippery one, bandied about by politicians with little respect for its subtleties. What, for example, is an average household?

Consider nine houses in Acacia Avenue.

- Four houses have 0 children
- One house has 1 child
- Three houses have 2 children
- One house has 15 children (OK, it's not a typical household)

What is the average number of children per household? You might remember that there are three common ways of expressing an average:

- The *mode*. This defines the average as the category that appears most often. In the case of Acacia Avenue, the most common number of children in a household is zero so this is the modal average. But it seems ridiculous to say that in Acacia Avenue the 'average' household has no children, since there are clearly lots of them running around in the street.
- The *median*. This says that the average is the middle value in the list if all the figures are put in order, from smallest to largest. In Acacia Avenue, the numbers of children in the nine households are: 0,0,0,0, 1, 2, 2, 2, 15. The middle, or median, value is 1. This seems odd too. How can the average household have one child when it is far more likely that a household has two children or none at all?
- That leaves us with the most popular form of average, the *mean*. This adds up all the values and divides them by the total number in the group. There are 22 children in the road and 9 houses, which means an average of roughly 2.4 children per household. Arguably this is the biggest nonsense of all, since *none* of the houses have exactly this many children, and only one house has more than this.

Still, *means* are the most popular form of average, including, as it happens, the one that is used to express the average income of the population. Average income is calculated by adding up all the incomes and dividing the result by the total number of people. In Britain, this figure is close to £30,000. Very few people earn exactly this amount, of course. In fact, as it happens, far more people earn less than this average income than more. This is because income is not evenly distributed. The majority earn less than £30,000, but a significant number earn between

£50,000 and £100,000, and there are many thousands who earn vast salaries up to the high millions. These big earners distort the average in the same way that the large family in Acacia Avenue distorted the average number of children per house.

This means that it is easy for a spin doctor working in opposition to make the voters unhappy with the government. 'I wonder how many people looking at this average income are thinking, "It's all right for the other half, but what about me?"' says the canny spokesperson, knowing full well that (a) well over half of the people watching are in 'the poor half' and, (b) in any case, whatever people are paid they always think they deserve more. It's a simple trick, but very effective.

But now for something spectacular. The conjuror David Copperfield has had a reputation for performing stunts on a large scale, but this is nothing to what a spin doctor can do. It's possible to increase the average wealth of two entire countries by moving just one person. Not convinced? Here's how.

Let's say the average (mean) income per person in Scotland is £29,000 per year, while the average income in England is £31,000 per year. (These are not far off the published figures.)

An Englishman, Wilf, on a salary of £30,000, is being transferred on the same pay from the London office to the Edinburgh office of his employer. Since Wilf's salary is lower than the average in England, his disappearance from the English statistics means that the average income in England will go up very slightly. Meanwhile, since his income is higher than the Scottish average, when he moves location the average income in Scotland will also increase slightly. So, by moving office, Wilf increased the average wealth in both countries.

This is beyond a mere conjuring trick: it seems to be positively miraculous. Yet, once again, the figures are absolutely true. It's just the conclusion that is false. The averages may have improved, but this is simply demonstrating the limitation of averages as a measure. The total wealth of England and Scotland has not changed with Wilf's moving, it has simply been distributed differently. But imagine what the spin doctors can do with this powerful tool!

Better than average

'I'm delighted to announce,' said the headmaster 'that this year, half of our pupils performed better than average. However, the other half of you are going to have to pull your socks up.'

In the 'common-sense' use of the word 'average', what the headmaster said was of course complete nonsense. Since the average is the midpoint, there will always be a half who perform worse than average, no matter how well they do.

Strictly speaking, however, this is only true if the average being referred to is the median. If the average referred to is the 'mean', then it is possible for more than half or less than half to be above average.

Missing the big picture

Another ploy of the good conjuror is to make you concentrate on a small part of what is going on so that you completely miss something else. Good patter will normally help with this. Consider, for example, this graph showing the decline in hospital waiting lists in a regional health authority:

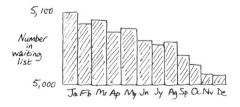

Impressive, isn't it? Sitting next to a photo of the chief executive with the caption 'We're making progress', it leaves an overriding impression that things are going very well. What the conjuror doesn't want you to look at too closely is the scale on the left-hand side of the graph. In fact, the actual decline in people waiting over six months has been about 100 out of a total of 5,000 – a minuscule 2 per cent. The graph looks rather different if the left-hand axis is shown all the way to zero:

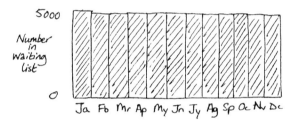

The so-called improvement in the waiting lists is so tiny that it's hardly worth mentioning. Once again something big has been made out of something small.

Selective presentation of the figures like this is so much the norm that a spin doctor would probably regard this as accurate reporting, but of course it is a presentation of numbers designed to convey a completely different impression from the truth.

Blind them with science

Finally, of course, there is the big flash, the mesmerising bit of chicanery that leaves the audience saying, 'Wow, I've no idea how they do that!'

One way to keep out prying eyes is to send out the message, 'We are so clever, it's not even worth trying to understand what we do.'

A standard way of doing this is to make simple things complicated, with the implication that complicated = sophisticated. The truth is, of course, that complicated often means no more than muddled thinking.

One of many such stories was that 'scientists' (whoever they are) had worked out a formula for the perfect football commentator. The formula, as merrily published in one newspaper, was as follows:

$$SQ = P - OL/2 + (LV \times 2) + Ra/2 + Rh + (T \times 1.5) - C/2$$

SQ stands for Speech Quality, and the other variables include Pitch, Loudness, Rhythm, Tone and so on.

It was all presented by the press in a slightly tongue-in-cheek way, since clearly this formula is complete gobbledygook. Only people involved in the research in question can evaluate whether it makes sense, and to everyone else it has no practical use. It is just a string of letters, but, because it is in the form of mathematics, it is supposed to be, by definition, a sign of cleverness and boffinhood; indeed it is little short of magic that can never be exposed.

A lot of maths is extremely difficult, but most of the maths needed for everyday life is not. Throughout this book we have tried to show that an understanding of maths can have all sorts of benefits: it can stimulate curiosity, it can answer those questions that bug us all the time, it can improve decision-making, and it can help to settle arguments. But perhaps the most important role of maths in everyday life is that it can help to prevent us from being conned, defrauded, misled and otherwise ripped off. There is nothing that spin doctors would like more than a generally innumerate society, so that we can be fed exactly the numbers they want to feed us.

With mathematics, it is possible to fight back.

APPENDIX

Proof that √2 is not a rational number

A number is called *rational* if it can be expressed as the ratio of two whole numbers. For example 2.75 is rational (11/4), as is 0.333333 … (= 1/3).

We can prove that √2 is not rational using the *reductio ad absurdum* approach (see page 196).

Let's start by supposing that √2 is a ratio of two whole numbers, a and b. Let's put that fraction in its simplest form, cancelling out any number that divides into the top and bottom of the fraction. So for example 39/13 can be reduced to 3/1.

In our fraction a/b, **a and b can't both be even**, because if they were, we could simplify the fraction by dividing a and b by 2 (for example 14/8 can be simplified to 7/4).

We want to see what happens if it's true that √2 = a/b

Square both sides: $a^2 / b^2 = 2$

Which means $a^2 = 2b^2$

Therefore a^2 is an even number, which means that:

a **must be even** i.e. a is twice another whole number, so a = 2c (where c is some other whole number)

$(2c)^2 = 2b^2$

$4c^2 = 2b^2$

Divide both sides by 2:

$2c^2 = b^2$ which means **b is even**.

But if a and b are both even, this contradicts our earlier assertion. The logical conclusion of this is that our original conjecture that √2 is a rational number must be wrong. So √2 is irrational.

Articles and papers

The Ghost's Vocabulary, Edward Dolnick
New Directions in Text Categorisation, Richard Forsyth
Literary and Linguistic computing, Richard Forsyth and
    David Holmes
The Development of Musical Tuning Systems, Peter A Frazer
The Weakest Link, John Haigh
The Rise and Fall of the Pyramid Schemes in Albania,
    Chris Jarvis
Rowling and 'Galbraith': an authorial analysis, Patrick Juola
The Mathematics of Diseases, Matt Keeling
The Power of One, Robert Matthews
Pythagorean Tuning and Medieval Polyphony,
    Margo Schulter
*The Best Known Packings of Equal Circles in the Unit
    Square*, E. Specht

Many of these articles can be found using online searches.

INDEX

Why Do Buses Come In Threes?

Why is it better to buy a lottery ticket on Saturday? Why do showers always seem to veer between too hot and too cold? And which classic puzzle was destroyed by Allied bombing in the war? These and many other questions are answered in this entertaining and highly informative book. *Why Do Buses Come in Threes?* is for anyone wanting to remind themselves – or discover for the first time – that maths is relevant to almost everything that we do. Dating, cooking, travelling by car, gambling and life-saving techniques all have links with intriguing mathematical problems that you will find explained here – including the odd coincidence of July 4th, the exponential growth of Australian rabbits and a surprising formula for running in the rain without getting wet. Whether you have a degree in astrophysics or haven't touched a maths problem since school, this book will change the way you view the world around you.

Rob Eastaway & Jeremy Wyndham

ISBN 9781911622277

100 Maddening Mindbending Puzzles

Puzzles have intrigued and entertained generations of children – and their parents – for over 2,000 years. Here is an irresistible assortment of 100 challenging puzzles, in a sumptuous new edition of this classic book. These brilliant brainteasers range from the neatly lateral to the downright perplexing. From chopping the chocolate to crossing the moat, mystifying matchsticks and a teasing typewriter, *100 Maddening Mindbending Puzzles* provides many hours of mind-stretching enjoyment for even the most agile brains. Guaranteed. If you consider yourself a master of logic, a devil with a crossword and a whizz at cracking codes, this is your chance to prove yourself a real smartypants and pit your wits against this devilish collection of games and puzzles.

Rob Eastaway & David Wells

ISBN 9781911622130